Weird and Wonderful
DINOSAUR FACTS

Written by Anne Rooney

Illustrated by Ro Ledesma

ARCTURUS

ARCTURUS

This edition published in 2026 by Arcturus Publishing Limited
26/27 Bickels Yard, 151–153 Bermondsey Street,
London SE1 3HA

Copyright © Arcturus Holdings Limited

All rights reserved. No part of this publication may be reproduced, stored in a retrieval system, or transmitted, in any form or by any means, electronic, mechanical, photocopying, recording, or otherwise, without written permission in accordance with the provisions of the Copyright Act 1956 (as amended). Any person or persons who do any unauthorized act in relation to this publication may be liable to criminal prosecution and civil claims for damages.

Writer: Anne Rooney
Illustrator: Ro Ledesma
Designer: Supriya Sahai
Editors: Lydia Halliday and Lucy Doncaster
Managing Editor: Joe Harris
Managing Designer: Rosie Bellwood-Moyler

ISBN: 978-1-3988-3338-8
CH011595NT
Supplier 29, Date 1125, PI 00011264

Printed in China

Authorised Representative:
Easy Access System Europe - Mustamäe tee 50, 10621 Tallinn, Estonia
gpsr.requests@easproject.com

ABOUT THIS BOOK

This weird and wonderful book is packed with extraordinarily epic facts—all about dinosaurs! You'll learn why some dinosaurs danced, how they communicated with each other, and what they liked to eat for dinner.

You can read this book from beginning to end, or you can flip through the pages until you find something that grabs your attention. There's also a pronunciation guide at the end, to help you sound out the names correctly.

So what are you waiting for? Open up and dig in—it's dino time!

SAUROPODS COULDN'T WEAR GLOVES

Huge sauropods like *Diplodocus* had no fingers, so gloves would have been useless to them! They had only one claw on each front foot (hand!) and it was on the thumb, facing inward, so was not easily seen in a side view. You often see them drawn with front feet like those of an elephant, but that's wrong.

BONY COLUMNS

Many sauropods didn't have separate fingers like you do. Instead, the fingers of their "hands" grew into a single column with the bones bunched together inside. Only the thumb kept its claw, and some sauropods didn't even have that. Some of the last and largest sauropods didn't have all their finger bones. Their fingers ended with stumps inside their column-hands, so they couldn't even wear mittens!

You can make a sauropod hand with your fist. Curve all your fingers inside and leave only your thumb out, then rest on your knuckles. Your thumb is too long, but otherwise the shape is about right.

HORSESHOE HANDPRINTS

With their fingers grouped together in a fleshy column, animals like *Diplodocus* left horseshoe-shaped handprints. We know from *Diplodocus* tracks—which are sometimes found fossilized—that their back feet had a very different shape to their "hands."

Something made a big impression.

MORE AT THE BACK

Sauropods had more claws at the back, often with three toe claws. Even these weren't quite what you might expect, though. They were arranged at a slant rather than sticking straight out. The dinosaur might have used these claws to scrape a hollow in the ground to lay its eggs in. (They didn't go in for fancy nest-building.)

Fast Facts

Name: *Diplodocus*
Size: 25 m (82 ft) long
Lived: 153 million years ago, western USA*
Ate: Leaves
*Most dinosaurs lived over a period of several million years; we give the middle of the range in these Fast Facts boxes.

SOME DINOSAURS WERE FLUFFY

People often assume that dinosaurs were all scaly like modern reptiles, such as snakes and lizards. But many dinosaurs had feathers—either full feathers like a bird, or fluffy down, or strange quills like a porcupine.

NO-FLY ZONE

Having feathers didn't mean dinosaurs flew around. Very few of them could fly. But flying isn't all that feathers are good for. If you've ever snuggled under a feather quilt, you'll know that feathers keep you warm. Feathers possibly first evolved to keep dinos warm, to keep their eggs warm in a nest, or maybe to attract a mate with their bright patterns.

FROM STRINGY STRANDS TO FLUFFY FEATHERS

Feathers developed from simple strands, like bristles or quills. Over many years, the bristles evolved into a bunch of strands and then developed a central "vane" from which the strands sprouted.

Fast Facts

Name: *Sinosauropteryx*
Size: 1 m (3.3 ft) long
Lived: 125 million years ago, China
Ate: Small reptiles and mammals

Sinosauropteryx, from China, had thin, hollow strands over its back and legs, but it definitely didn't fly.

FEATHERY CROCODILES?

Feathers developed before dinosaurs. Scientists now think pterosaurs had a fuzzy coat, so the ancestors of both pterosaurs and dinosaurs possibly had some kind of feathers. Crocodiles evolved from the same group of reptiles, so they might once have had feathers, too!

DINOSAURS AREN'T ALL DEAD...

... because birds are dinosaurs! Birds are the only surviving type of dinosaur, since all the others went extinct (died out) 66 million years ago when a giant rock from space crashed into Earth.

FROM ARMS TO WINGS

Birds evolved from a type of dinosaur named theropods—one of the three basic dinosaur designs. Theropods were usually speedy animals that ran on two legs and ate meat. *T. rex* and *Velociraptor* were both types of theropod.

In China, several small theropods have been found that had feathers and were starting to look like birds. Some of these probably had feathery arms which they flapped, and some might even have glided on outstretched feathery arms that were starting to work like wings.

HALFWAY THERE

Archaeopteryx lived in Germany 150 million years ago. It had feathers, wings, and claws, like a bird, but it had teeth (birds don't), and bones in its tail (birds don't). Starting 160 million years ago, lots of similar nearly-birds lived in China. The earliest ones couldn't fly, but by 125 million years ago there were flying dinosaurs, and the first birds appeared.

Fast Facts

Name: *Archaeopteryx*
Size: 50 cm (19 in)
Lived: 160 million years ago, Germany
Ate: Small animals

Watch out—low-flying zone!

GOING BACKWARD

Not all birds fly now. Ostriches and emus have given up flying and look quite reptilian with their scaly feet and bald heads. They can run and kick, just like the theropod dinosaurs. And penguins have swapped flying for swimming, changing their shape to become streamlined, and their wings to work as flippers.

PARASAUROLOPHUS HAD ITS OWN IN-BUILT TRUMPET

It had a long, hollow crest on top of its head, which it could probably have used to make trumpeting noises. Imagine the woods of North America filled with the sounds made by herds of *Parasaurolophus*!

AIRWAVES THROUGH THE AIRWAYS

When it wanted to make a noise, *Parasaurolophus* took a breath then blew it out through the long tubes of its crest. A channel went from each nostril up to the end of the crest, where it turned around and doubled back on itself.

One scientist even made a plastic model of the crest of a *Parasaurolophus* and played it as a musical instrument. He found it was very much like an instrument called a crumhorn, and when he played it, it sounded like a tuba.

DIFFERENT HEAD, DIFFERENT NOISE

There are three different species (types) of *Parasaurolophus*, with two slightly different styles of crest. *Parasaurolophus* had a long crest which made a trumpeting noise with a low tone, while others with a shorter crest made a higher-pitched sound. The different sounds would mean the dinosaurs could easily identify others of the same species. Little baby *Parasaurolophus* made a much higher-pitched sound.

Fast Facts

Name: *Parasaurolophus*
Size: 9.5 m (31 ft)
Lived: 75 million years ago, northwest USA and western Canada
Ate: Leaves, twigs, pine needles

TALK TO ME?

We don't know what *Parasaurolophus* used its trumpet for. It might have been to attract a mate, or to warn others of any danger it spotted. Or maybe they just liked to chat!

You're playing my tune.

DINOSAURS ARE MADE OF ROCK

Well, they are now. They weren't when they were alive! Dinosaur scientists find out about dinosaurs by looking at fossils, which are traces and parts of dinosaurs that have turned to stone over millions of years.

DEAD AND WET

For a fossil to form, the body of a dinosaur had to be quickly covered after it died. It could be covered with sand, sediment (material deposited by water, wind, or ice), or volcanic ash. This had to happen before the dead dinosaur was eaten, or destroyed by the weather, or rotted away. Then the body had to be covered by water. If it fell in a river or its body was washed away by a flood, that was easy. If it died in the middle of a desert, it probably wouldn't make it as a fossil.

IN AND OUT

Bones have lots of tiny holes in them—even your bones do! When a bone lies in wet sediment, water seeps into it and fills the holes. Slowly, minerals dissolved in the water are deposited in the holes, eventually filling them in. Sometimes, the minerals that make up the bones or other body part are changed to other minerals through chemical reactions. The bones still have the same shape, but are made entirely of stone.

DEEP IN THE ROCKS

Sediment falling over the body buries it deeper and deeper. As pressure builds up, the sediment turns to rock, trapping the animal's changed skeleton inside. If the rock is revealed, perhaps in a landslip or an earthquake, the fossils come to the surface, and we can find them.

DINOSAURS DANCED THE NIGHT AWAY

Or perhaps the day. It could have been part of a mating ritual in which dinosaurs showed off to each other. Imagine a lot of dinosaurs strutting their stuff to attract a partner—just like a disco, but without the music.

BEHAVE LIKE A BIRD

Theropod dinosaurs were the ancestors of modern birds. Many of them had feathers, some sat on their eggs in nests, and they probably behaved like birds in some other ways, too. Some types of bird carry out fancy mating rituals—the males perform in front of the females in a parade called a "lek." Other types of bird target a particular female and try to persuade her that they are the perfect partner.

You're a dino dazzler.

MOONWALKING AND TAP DANCING

Birds have lots of ways of dancing to impress each other. Dinosaurs might have done any or all of these—or something entirely different. For example, a male red-capped manakin moonwalks along a branch, while a magnificent riflebird stretches his wings out and bobs his head from side to side while jumping. And a blue-footed booby lifts one leg after the other, showing off his feet, while a Jackson's widowbird jumps up and down enthusiastically.

Perhaps a *Velociraptor* tap-danced like the stamping sharp-tailed grouse, or a *T. rex* jumped thuddingly up and down, making the ground shake? We know very little about how dinosaurs behaved—actions don't form fossils! But looking at what their descendants do can show us some of the things they might have done.

T. REX DIDN'T ROAR

If you've ever seen a dinosaur movie, it probably showed big, fierce dinosaurs like *T. rex* roaring in a scary, ferocious way. But *T. rex* and other dinosaurs didn't have the right type of throats to roar.

GRUMBLING AND BOOMING

Large dinosaurs might have made low booming or grumbling sounds with their mouths closed. This is what large birds like ostriches do, and also crocodiles. Since crocodiles, birds, and dinosaurs have all evolved from the same group of earlier animals, scientists think they might all make similar sounds.

CHEEPING, TWEETING, AND COOING

Small dinosaurs, and particularly young dinosaurs, might have made much higher-pitched noises, more like the chirping of small birds or even the cooing of pigeons. Dinosaur ear bones seem to have been suited to hearing different pitches, so they might have been listening out for their babies at the higher end of their hearing range and listening to other adults at the lower end.

Fast Facts

Name: *Tyrannosaurus rex*
Size: 12.5 m (41 ft)
Lived: 68–66 million years ago, northwest USA and western Canada
Ate: Other dinosaurs

TOO LOW TO HEAR

It's also possible that some dinosaurs made sounds of such a low pitch that we wouldn't have been able to hear them. Infrasound—sound too low for us to hear—is made by some large animals today, including elephants, rhinos, and alligators. Perhaps a quiet *T. rex* would have been even more scary than a roaring one—silent and deadly.

T. REX LIVED CLOSER IN TIME TO A PHONE THAN TO STEGOSAURUS

You've probably seen pictures of *T. rex* in the same landscape as *Stegosaurus*. But in fact, they never met. *T. rex* lived 66 million years ago, while *Stegosaurus* lived 145 million years ago. The steggies died out about 80 million years before *T. rex* first stomped across North America.

"DINOSAUR" ISN'T AN INSULT

Overall, dinosaurs existed for about 180 million years. When you hear people criticizing things they think are unsuccessful and will quickly die out, calling them "dinosaurs," they're totally wrong. Humans like us—*Homo sapiens*—have been around for less than half a million years, so the dinosaurs are winning in the survival stakes. They lived for more than 360 times as long as we've managed so far.

Fast Facts

Name: *Stegosaurus*
Size: 7 m (23 ft)
Lived: 150 million years ago, western USA and Portugal
Ate: Plants

THEY DIDN'T KNOW THEY WERE DINOSAURS!

Of course, dinosaurs weren't all the same. They're a group containing lots of different species, from tiny animals no bigger than a chicken to the largest animals that have ever walked on Earth. We group them together, but dinosaurs wouldn't have seen any similarity between different types. Dinosaurs often ate other dinosaurs, just as many mammals eat other mammals, fish eat other fish, and birds eat other birds. The grouping that gives a bunch of reptiles the label "dinosaur" is very much a bit of human thinking.

VELOCIRAPTOR AND T. REX NEVER MET

Dinosaurs not only lived at different times, but they also lived in different places all over the world, including on land that is now Antarctica. This meant that many species never met. For example, *Velociraptor* lived both 3 million years earlier than *T. rex* and thousands of miles away.

A LONG TREK

T. rex and *Velociraptor* lived near the end of the age of the dinosaurs. By that time, the map of the world looked similar to how it is now. It wasn't quite the same though—there was sea between North and South America and much of Europe was under water. It was also warmer and sea levels were higher, so there was less land and more ocean.

AN IMPOSSIBLE ROUTE

There was no land route connecting the home territories of *T. rex* and *Velociraptor*, so there was no way they could have met, even if they had been alive at the same time. But in earlier times, all the land on Earth was joined into a single, huge supercontinent, named Pangea. If the two dinosaurs had lived 200 million years ago, they could have trekked between Texas and Mongolia.

Fast Facts

Name: *Velociraptor*
Size: 1.5–2 m (5–6.8 ft) long
Lived: 73 million years ago, Mongolia
Ate: Other dinosaurs

DINOS EVERYWHERE

Because the land was once all joined, early dinosaurs could walk anywhere. Over thousands or millions of years, they spread out over the whole Earth. When Pangea started to split up, the dinosaurs were already spread out, so nowhere was dino-free.

Are we there yet?

SOME DINOSAURS ATE STONES ...

... not for nourishment, but to help process food in their stomachs. (They did eat other things as well as stones.) Today, some birds, as well as some reptiles, amphibians, and even worms also swallow stones. The stones are called "gastroliths," which means "stomach stones."

TOUGH COOKIES

Plants are made of a very tough material called cellulose that is hard to break down. Animals that eat plants need a gut that can cope with the task. That's as true for the dinosaurs as it is now. Many plant-eating dinosaurs were very large, so they had to eat a lot to fuel their huge bodies. They swallowed mouthfuls of leaves and twigs without chewing. If they'd stopped to chew, they wouldn't have had time to eat enough food. They didn't even have any chewing teeth. All the processing had to be done in their guts, and that's where the stones come in.

Fast Facts

Name: *Cedarosaurus*
Size: 15 m (49 ft) long
Lived: 136 million years ago, midwestern USA
Ate: Plants

CHURNING AND TURNING

In an animal's gut, muscles work to churn the food, turning and squashing it so that stomach acid can work on it more quickly. Throwing in some stones helps the process because they grind the food into smaller particles as the stomach tumbles them around.

SMOOTH AND SHINY

If you've ever seen pebbles on a beach, you'll know that they're rounded and smooth. Tumbling around against other stones eventually wears away all the hard edges. The same happens inside an animal's stomach. Smooth stones clustered in the gut area of a fossilized dinosaur are generally gastroliths. There can be several kilograms of stones in the gut area of a large dinosaur, some of them quite big!

T. REX TEENAGERS HAD A GIGANTIC GROWTH SPURT

Human parents are often astonished at how much a teenager can eat, but they'd be even more surprised if they were raising a family of *T. rex*. A teenage *T. rex* had a huge growth spurt, putting on 2.3–3 kg (5–6.5 lb) a day!

LOTS OF GROWING TO DO

An adult *T. rex* could weigh 7,300 kg (16,100 lb) and be 12.8 m (42 ft) long. It had to get to that size from a small hatchling that fitted, curled up, in an egg just 43 cm (17 in) long. The baby would be about the size of a collie dog, and would have to grow to 13 times this size to become a full-sized adult.

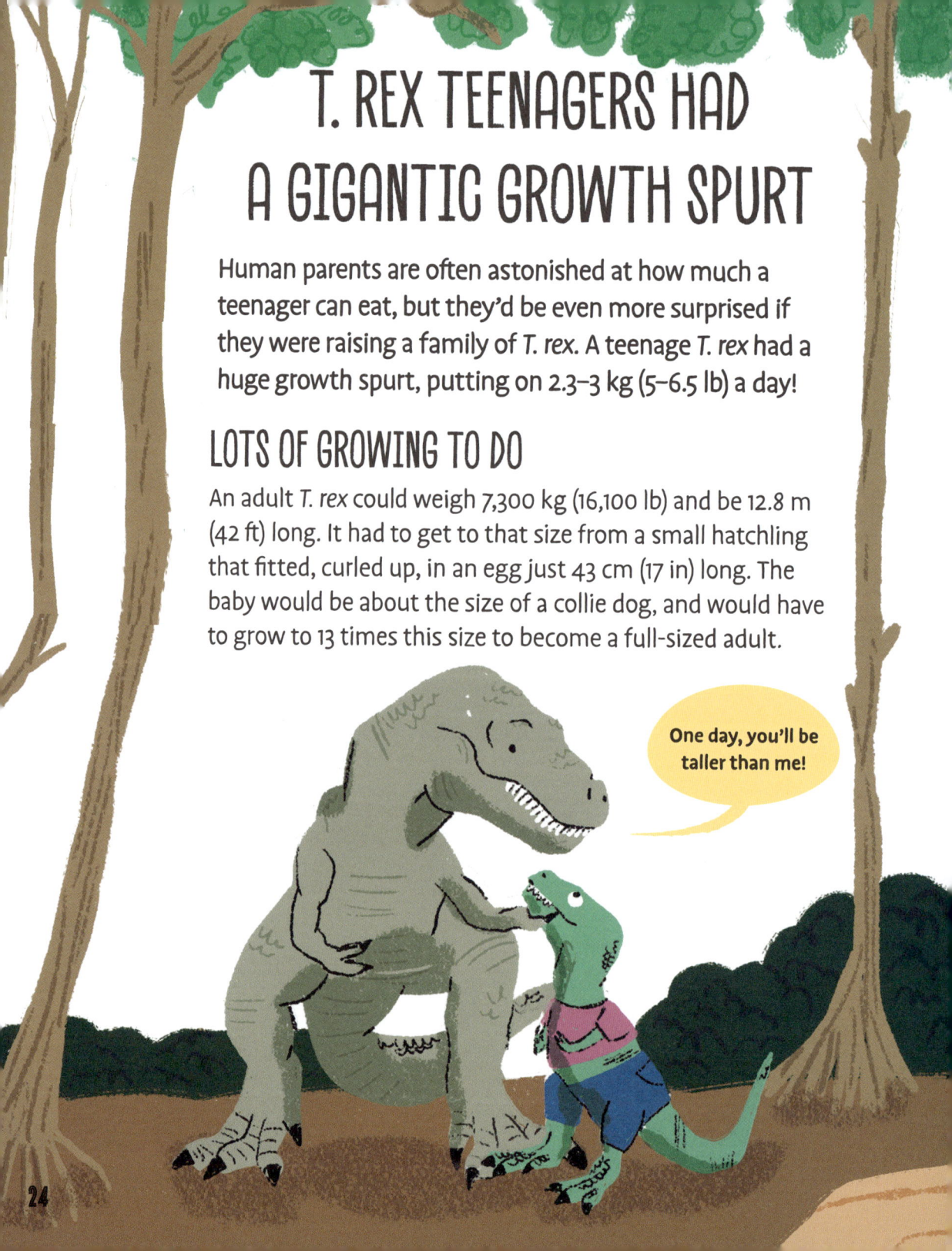

One day, you'll be taller than me!

STARTING SMALL

Though grown-up *T. rex* could tackle pretty much any animal in their environment, the babies probably ate insects and other tiny creatures. They ate larger things as they grew, and so could soon put on weight.

Fast Facts

Name: *Tyrannosaurus rex*
Size: 12.5 m (41 ft)
Lived: 68–66 million years ago, northwest USA and western Canada
Ate: Other dinosaurs

IN THE BONES

Dinosaur scientists know the proportion of, say, a dinosaur's thigh bone to the rest of its body, so they can work out the overall size from just a few bones. They can also work out the age of a dinosaur by counting rings in its bones. By comparing different sizes and ages of the same dinosaur type, they can calculate how quickly they grew at different stages.

DINOSAURS COULD HAVE SEEN DINOSAUR FOSSILS

We know about dinosaurs because we have found fossils of them. Fossils form over millions of years, but dinosaurs of different types lived on Earth for such a long time that later dinosaurs could easily have come across fossils of earlier dinosaurs.

DEAD BUT NOT GONE

Fossils can form when an animal's body falls into water or mud and is covered by sediment (such as mud and sand), and then water. Chemicals in the water react with the minerals in the animal's bones, claws, teeth, and other hard parts, slowly changing them to stone. Over many thousands or millions of years, the parts become fossils.

FOSSILIZED HOLES

As well as body parts, sometimes empty shapes fossilize! These can be footprints, or the spaces where soft body parts such as feathers and flesh once lay, or even burrows. Sediment fills in the gaps and hardens over time, making a fossil in the shape of what's gone. These are called "trace fossils."

Fast Facts

Name: Fossils
Size: Range from microscopic microbe fossils to giant sauropods 26 m (85 ft) long
Age: Dinosaur fossils range from 240 million years ago to 66 million years ago

These butt prints will baffle scientists.

A LONG TIME DEAD

Dinosaurs like *T. rex* or *Triceratops* that lived 66 million years ago might well have seen fossils of earlier dinosaurs, such as *Diplodocus*, that lived millions of years earlier in the same place. The first dinosaurs evolved around 240 million years ago, so there was plenty of time for fossils of them to form before the later dinosaurs came along. Even *Diplodocus* could have seen fossils of older dinosaurs. Dinosaurs probably wouldn't have recognized any fossils they saw, though.

VELOCIRAPTOR HAD A WISHBONE— JUST LIKE A CHICKEN

If you've ever had a roast chicken, you might have pulled the wishbone in half. This is a V-shaped bone that scientists call the "furcula." It's a single bone made from two fused clavicles, or collar bones. In some countries, pulling the wishbone apart traditionally grants a wish to the person who gains the largest portion.

A BONE FOR FLYING

You have two clavicles across the top of your chest. In a bird, these are fused into one bone, to make the chest area strong enough to support the muscles needed for flying. It also acts like a spring. The ends are stretched apart when the bird raises its wings, and they snap back together on the downward stroke. But a *Velociraptor* couldn't fly, so what's it doing with a wishbone?

That Velociraptor was delicious—but I could have done with some fries!

Fast Facts

Name: *Velociraptor*
Size: 1.5–2 m (5–6.8 ft) long
Lived: 73 million years ago, Mongolia
Ate: Other dinosaurs

GOING NOWHERE

Velociraptor wasn't the only non-flying dinosaur to have a furcula. Some other theropods (like *T. rex*) also had one. Some scientists think the furcula developed independently in animals like *Velociraptor* and in bird ancestors. Others think it developed for a different purpose and that birds later started to use it to help with flying.

Perhaps it was useful for *Velociraptor* and others to be able to snap their arms down and forward with rapid force, so the "spring-loaded" mechanism suited them. And then birds found it helpful when flying. Whatever the answer, you'd get a serious-sized wish from the wishbone of a dinosaur!

DINOSAURS IN CALIFORNIA COULD NEVER VISIT NEW YORK

And not just because it wasn't built! You'd think huge dinosaurs, given enough time, could walk from one side of North America to the other. But North America was split top-to-bottom by a seaway that kept dinosaurs in the east separate from those in the west. Most of the American dinosaur fossils we have are from the west.

STRANDED

The "Western Interior Seaway" was a shallow sea that split North America. Dinosaurs on one side could never get across to the other side. Not only that, they couldn't go anywhere else either, since they were surrounded by different bodies of water.

NOT LIKE THE OLD DAYS

It hadn't always been like this. Once, all the land formed a single, huge, supercontinent named Pangea. While the land was linked, dinosaurs walked around and lived in different places. But about 250 million years ago, Pangea began to split up. Oceans opened and lands drifted apart. Dinosaurs trapped on blocks of land began to evolve separately, adapting to conditions in their own particular region.

I'll travel abroad next year.

TEEMING SEA

The Western Interior Seaway might have been a nuisance for dinosaurs, but it was a good place for animals that lived in water, like plesiosaurs, fish, and sharks. Their fossils are now found inland because the land that was once sea is now part of the American mainland.

THERE WAS ONCE A SHORTAGE OF SAUROPODS

Most of the time, there was a good mix of dinosaurs in many places. There were big ones, small ones, middle-sized ones, and a few truly huge ones. The huge ones were the sauropods, like *Diplodocus* and *Brontosaurus*. But for 30 million years, there were no sauropods in North America or Europe (as far as we know).

SAUROPOD, SAUROPOD, NO SAUROPOD, SAUROPOD ...

The name for the period for which no sauropod fossils have been found is the "sauropod hiatus." It lasted 25–30 million years, from 104 to 70 million years ago. Then *Alamosaurus* arrived in North America just in time to be killed off by an asteroid strike 66 million years ago. Bad timing! There was also a worldwide shortage of sauropods early in the Cretaceous period, 145 million years ago. This came just after the sauropods' time of glory—the Jurassic period.

Scientists wonder whether an extinction event struck the sauropods at the end of the Jurassic period, wiping them out in North America and Europe. *Alamosaurus* might have later arrived in North America from South America, where there were still plenty of sauropods.

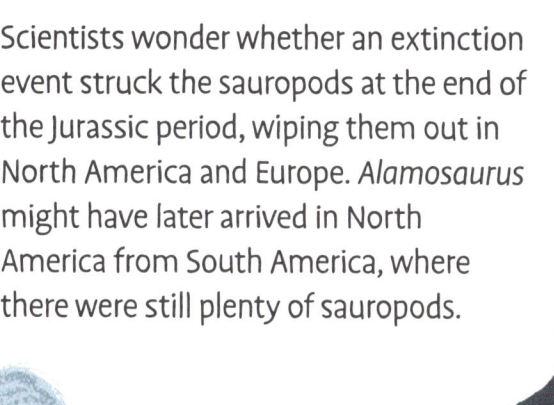

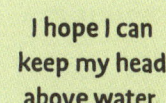
I hope I can keep my head above water.

HIDDEN SAUROPODS?

It's possible there were sauropods during the hiatus after all, but we haven't found them yet. One idea is that sauropods didn't like the beach—they preferred to live properly inland. As North America was split by the Western Interior Seaway (see page 30), there was more beach and less space for landlubber sauropods. They would have moved away from the edges, and that also means they moved away from the areas where fossil hunters have found most dinosaurs.

Fast Facts

Name: *Alamosaurus*
Size: 26 m (85 ft) long
Lived: 68 million years ago, southwest USA
Ate: Plants

THE DINOSAURS DIED ON A SPRING DAY

All dinosaurs, except birds, died out 66 million years ago in just a few days or weeks. It started on a spring day when a rock from space, measuring 10 km (6.2 mi) across, slammed into the coast of Mexico at 100,000 km per hour (62,100 mph). From looking at the growth patterns in fish fossils and insect damage to leaves, scientists worked out that the asteroid struck during spring in the northern hemisphere.

WORLDWIDE DEVASTATION

The asteroid made a crater 180 km (112 mi) across and 20 km (12 mi) deep. The impact hurled a huge amount of rock and dust into the air, causing a vast tsunami (flooding waves) up to 1.6 km (1 mi) high and winds of up to 1,000 km per hour (620 mph). Everything within 1,500 km (930 mi) of the crash site died instantly. Shock waves ran through Earth, causing earthquakes and tsunamis far away. Falling particles of rock heated the air so much that wildfires began.

Fast Facts

Name: Chicxulub asteroid
Size: 10 km (6.2 mi) across
Struck: 66 million years ago, Bay of Mexico
Killed: Three-quarters of all species on Earth

A LONG, HARD WINTER

Dust and gas in the air blocked sunlight, so Earth became dark and cold. Global temperatures dropped by 25 °C (45 °F). The planet was semi-dark for ten years. Plants couldn't grow, so plant-eaters died, then the meat-eaters died. The temperature stayed below freezing for years, and the ocean took centuries to warm up again.

DINOSAURS HAD FLEAS ...

... and big dinosaurs had big fleas! Biting parasites like fleas, ticks, and mosquitoes have been around for a long time. A fossilized Jurassic bug, nearly 2.5 cm (1 in) long, even seems to have bitten dinosaurs!

I wish someone would invent the flea collar!

BIG AND BRUTAL

The ancient bugs were about ten times the size of a modern dog flea, and had a vicious serrated mouth part for cutting through thick dinosaur skin. Their long, hooked legs could have gripped scales or feathers while the insect fed on blood through the skin. They looked more like ticks than fleas, and didn't have the special legs for jumping that are familiar in cat and dog fleas, so they might not have been closely related to modern fleas.

BITTEN INSIDE AND OUT

Fleas and ticks live on the outside of an animal. In 2020, a scientist found the first parasite inside a dinosaur. It was a microorganism (a type of microscopic blood parasite), discovered inside the bones of a giant sauropod.

Fast Facts

Name: Jurassic pseudo-flea
Size: 1.7 cm (0.7 in) long
Lived: 145 million years ago, China
Ate: Blood

FEATHERS, BUGS, BEAKS, AND TEETH

Feathers provide a great home for parasites. The bugs can stay snug and protected, and are next to the skin so they can easily feed on blood. Many modern birds have an overhanging upper bill, which helps them to groom their feathers, picking out parasites. Many feathered dinosaurs and early birds had teeth, and a lot had smaller teeth at the front of their mouths. Some scientists suggest they might have used these to pick bugs from their feathers!

Dinosaurs come in such large portions!

SAVAGE RAPTORS WERE FUSSY ABOUT THEIR NAILS

Dinosaurs like *Velociraptor* and *Dakotaraptor* were speedy theropods that ran on two feet to catch their prey. They were very careful to take good care of their special toenails by holding them off the ground when running.

"KILLING CLAWS"

Raptor dinosaurs had a huge, curved claw on each back foot. When running or walking, they held this up off the ground to protect it (and stop them tripping over it). The claw came into its own when they made a kill. Scientists once thought that raptors used their "killing claw" to rip open the body of an animal they had caught. But now it seems they didn't do that. They used the claw to hold down their victim so they could eat it, much like the way modern birds such as owls hold their dinner.

Fast Facts

Name: *Dakotaraptor*
Size: 5.5 m (18 ft) long
Lived: 66 million years ago, western USA
Ate: Other dinosaurs

SCARY IN THE DARK

Raptor dinosaurs probably hunted animals that were smaller than themselves. *Velociraptor* lived in Mongolia 70 million years ago and probably preyed on small mammals, and perhaps lizards and baby dinosaurs too.

Do you ever get the feeling that you're being watched?

Scientists think *Velociraptor* hunted by night, as the structure of its skull suggests it had large eyes. Most nocturnal animals (those active at night) have large eyes to gather as much light as possible. Eating small mammals, hunting by night, and using its large claws to hold down its prey ... *Velociraptor* sounds a bit like a prehistoric owl! But owls aren't fussy about their toenails.

WE DON'T KNOW WHAT MOST DINOSAURS LOOKED LIKE

All we have of most dinosaurs is a few fossilized bones and teeth. We don't know what they looked like in the flesh, whether they had bright feathers or scales, whether they were fat or thin, nor do we know about anything else that lay over the bones.

BITS AND PIECES

Most dinosaur fossils are incomplete—just a few bones, teeth, and claws, usually jumbled up or spread over a wide area. There's rarely a whole dinosaur in one place, and it's very rare for it to be all together in the right order. It takes a lot of work for experts to figure out which type of dinosaur the fossils belong to, or if it's an entirely new type of dinosaur.

Bones such as ribs, vertebrae (links of backbone), and leg and arm bones all have distinctive shapes that are easy to recognize, even across different types of animal. So, it's fairly easy to work out what type the bone is—even if it's only a small part.

Fast Facts

Name: *Therizinosaurus*
Size: 10 m (33 ft)
Lived: 70 million years ago, Mongolia
Ate: Plants

PIECING IT TOGETHER

When bits are missing, scientists work out from similar types of dinosaur what the missing parts probably looked like. They don't just show drawings with gaps. If they did, *Diplodocus* would be shown without a head, as no complete skull has ever been found! Instead, a *Diplodocus* you see in a museum or drawing has the head of a related dinosaur that is probably similar to the *Diplodocus*. For *Therizinosaurus*, only the arms and a few toes and ribs survive, but there are plenty of pictures showing it complete.

PRETTY UBIRAJARA WAS DECORATED WITH RIBBONS

An odd little dinosaur from Brazil had pairs of stiff, ribbon-like decorations sticking out of its shoulders, and was covered in a mane of long feathery "hair." About the size of a chicken, *Ubirajara* lived 110 million years ago.

FINGERNAILS OR RIBBONS?

Ubirajara's "ribbons" were quite stiff and probably made of keratin, which is the same material that your fingernails and hair are made from. Which would you rather have, fingernails or shoulder ribbons? Fingernails are useful, but shoulder ribbons are spectacular!

Fast Facts

Name: *Ubirajara*
Size: 1.5 m (5 ft) long
Lived: 113 million years ago, Brazil
Ate: Small animals

Ubirajara possibly used its ribbons in a display to attract a mate, and it might be that only males had them. They could also have helped it look bigger and maybe scary (or just strange) to ward off slightly bigger dinos. It wouldn't have been easy or comfortable to grab *Ubirajara* from the front and then get a mouthful of ribbons.

LOOK AT ME!

Lots of animals have structures or patterns to attract mates. Many male birds have bright, long, or bizarre-looking feathers that they parade to females they want to impress. Sometimes these feathers must make life quite hard for the bird. Imagine being a male peacock and having to drag that tail around just so that you can spread it out occasionally to look impressive. Since birds evolved from dinosaurs, it's possible that these dinosaurs had to put up with the hassle of wide shoulders to look cool.

DINOSAURS LIVED IN ANTARCTICA

Dinosaur fossils have been found on every continent, even Antarctica. Today, Antarctica is a frozen wasteland with no large land animals at all. But in the age of the dinosaurs, it was a very different place.

CHANGING PLACES

The slabs of rock that make up land move slowly around Earth over millions of years. Sometimes they're grouped together into a single supercontinent, and sometimes they're spread around, separated by oceans. As they drift, landmasses are sometimes in warmer parts and sometimes in colder parts. The Earth has also been warmer and colder.

Fast Facts

Name: *Antarctopelta*
Size: 4 m (13 ft)
Lived: 78 million years ago, Antarctica
Ate: Plants

CLIMATE CHANGE

In the Cretaceous period, 145–66 million years ago, the whole world was warmer than it is now. There was more carbon dioxide in the atmosphere, sea levels were 100 m (330 ft) higher, and there was no permanent ice at the South Pole.

ALL THE BETTER TO SEE YOU WITH ...

Although it wasn't cold near the South Pole, the winters were still long and dark. Near both Poles, days in summer are long, with little or no darkness, but in winter, days are short and never brighter than twilight. Dinosaurs in Antarctica, such as *Antarctopelta,* probably had large eyes to gather as much light as possible, and maybe hibernated in the winter, like some modern reptiles do. Or they might have moved away, to a place where the days were lighter.

HUGE DINOSAURS CAME FROM EGGS THE SIZE OF A SOCCER BALL

All dinosaurs hatched from eggs, but even the very biggest didn't lay large eggs. In fact, they laid tiny eggs for their size. Huge sauropods that grew to 37 m (121 ft) long laid eggs only 15 cm (6 in) across. Their babies had a lot of growing to do!

GAMBLING ON BABIES

Sauropods laid spherical (ball-shaped) eggs, while some other dinosaurs laid long, thin eggs, but they were all small. Some animals raise one or two babies and put a lot of effort into them, and others produce lots of babies and hope some survive. Sauropods seem to fall into the second group. They laid a lot of eggs, possibly making several nests a year with lots in each, and hoped for the best.

GOOD MOTHERS?

Sauropods were too large to sit on their eggs without crushing them, but some smaller dinosaurs, such as Oviraptor, sat on their nests. This not only kept their eggs warm but protected from hungry egg-eaters.

His father was a *Giganotosaurus*.

LARGE AND SMALL EGGS

There's a limit to how big an egg can be and still work. Part of the problem is that it has to come from the mother's body. It takes a lot of energy and material from food to make and lay eggs, and bigger eggs require more than small eggs. To lay lots of eggs, a dinosaur had to make them quite small. In addition, a larger egg needs a thicker shell to protect it, and that makes the egg heavier and even more difficult to make. The biggest egg of any animal is 30 cm (12 in) long, laid by the extinct elephant bird from Madagascar.

Fast Facts

Name: *Oviraptor*
Size: 2 m (6.6 ft)
Lived: 75 million years ago, Mongolia
Ate: Shellfish, mollusks, hard fruit

SOME DINOSAURS GLOWED IN ULTRAVIOLET

They would have looked good in disco lighting. Ultraviolet (UV) is just beyond visible light in the spectrum. We can't see in ultraviolet, but many types of animals can. If you go to a disco with UV lights, some white items glow brightly—they "fluoresce."

FLUORESCENT CREATURES

Objects that are fluorescent absorb UV and then emit it at a different wavelength. Puffins have fluorescent patches on their beaks, and budgerigars have patches of fluorescent feathers on their heads. Budgies use these in displays when they want to attract a mate. Lots of frogs and salamanders are also fluorescent, glowing green under blue light, and the ocean has many fluorescent creatures.

Fossil hunting is light work.

GLOWING DINOSAUR

A fossil of the Canadian dinosaur *Borealopelta*, found in 2011, partly glows in UV. The dinosaur had two spikes 60 cm (24 in) long growing from its shoulders. Even 110 million years after its death, the tips of the spikes are fluorescent. The spikes have a core of bone and then an outer covering of keratin that made them even longer.

Fast Facts

Name: *Borealopelta*
Size: 5.5 m (18 ft) long
Lived: 105 million years ago, Canada
Ate: Plants

The dinosaur had other plates, knobs, and spikes with a keratin covering, too. Most keratin was found on structures near the front of the animal. The bony bits probably helped protect it from predators and also made it appear as a big, fearsome opponent when it was competing for mates. It would have looked like a good, healthy prospect to any mate it was trying to attract. It might have helped its chances if it could have shown that it glowed at the disco, too!

EARLY DINOSAURS WERE HORSE-SIZED

We think of dinosaurs as being massive, but not all of them were. The first ones were smaller—it was only in the Jurassic period that some grew huge.

A CLEAN SLATE

The dinosaurs emerged in a world wiped clean by disaster. Around 252 million years ago, the worst-ever mass extinction event wiped out 95 percent of all species. Named the "Great Dying," it was caused by huge volcanic eruptions that continued for hundreds of thousands of years. These poured out carbon dioxide and other gases, warming the climate until it was fatally overheated.

A couple of million years afterward, new animals emerged, such as *Prorotodactylus,* which lived in Poland 250 million years ago. The size of a cat, it's the earliest known "dinosauromorph," which means dinosaur-shaped animal. It wasn't a true dinosaur, but this type of animal was the ancestor of the dinosaurs that followed.

PICK YOUR BODY SHAPE

Dinosaurs evolved from the dinosauromorphs over millions of years. The first true dinosaurs we know of lived in South America 230 million years ago. At first, they all looked rather similar, and they were no taller than a horse. Over time, they separated into three groups with different body shapes.

Fast Facts

Name: *Herrerasaurus*
Size: 6 m (20 ft) long
Lived: 230 million years ago, Argentina
Ate: Small animals

These later became three types of dinosaur—the theropods, sauropods, and ornithischians. Theropods were meat-eaters that ran quickly on their hind legs. Sauropods were slow plant-eaters that walked on four legs and eventually grew to enormous sizes. Ornithischians were all the rest—plant-eaters with beaks that ranged from stumpy ankylosaurs to herds of nimble hadrosaurs.

DINOSAURS WON A TWO-HORSE RACE TO RULE THE WORLD

When the first dinosaurs emerged around 230 million years ago, they weren't the only kids on the block. They had to compete with some scary crocodile-like creatures. It took them 30 million years to win the contest.

FIERCE COMPETITION

Starting with a nearly empty world after the Great Dying (see page 50), other animals evolved alongside the dinosaurs. Their biggest rivals were pseudosuchians ("SOO-doh-SOO-Kee-uns"). They were quite like crocodiles, but came in all shapes and sizes. Some ran fast on two legs, some ate plants, some grew huge and had teeth like steak knives, and some might even have climbed trees. They were vicious, successful, and soon they were everywhere.

Dinosaurs, meanwhile, were still quite small—no bigger than a horse, remember (see pages 50–51). They weren't as chunky and powerful as the crocodilians, and lived only in warm, wet places, not in dry deserts. The pseudosuchians, which could live anywhere, were definitely winning.

CHANGE AGAIN

Things don't remain stable on Earth for long. Around 200 million years ago, there was another volcanic catastrophe. As the supercontinent Pangea split apart, lava poured from the gap that would eventually become the Atlantic Ocean.

> **Fast Facts**
> **Name**: *Eoraptor*
> **Size**: 1.4 m (4.5 ft) long
> **Lived**: 230 million years ago, Argentina
> **Ate**: Plants and small animals

Can't catch me!

Gas and dust blocked sunlight and the temperature plunged. It was a lucky break for the dinosaurs, but unlucky for their rival pseudosuchians. Most of the crocodilians died out, leaving just a few ancestors of modern crocodiles. At last, the dinosaurs had the stage to themselves. They soon grew and spread to fill it, becoming huge and ruling every continent for the next 145 million years.

THERIZINOSAURUS HAD CLAWS LIKE DAGGERS

It had the longest claws of any animal that has ever lived on Earth. But even though its vicious-looking claws were more than 50 cm (20 in) long, *Therizinosaurus* only ate plants.

TERRIBLE CLAWS, TINY JAWS

The claws look as though they would be perfect for ripping into prey, but dinosaurs that had long claws like these on their hands generally used them to pull plants toward them. *Therizinosaurus* could hook a claw around a branch or through a bunch of leaves and pull it to its mouth. No fossil skull of *Therizinosaurus* has been found, but scientists know from similar animals that it likely had quite weak jaws and a hard beak for biting through bits of tree. It was also tall, so it could reach up high.

NO FIGHTING

The claws might have made *Therizinosaurus* look scary, and that might have been enough—with its large size—to put other animals off attacking it. But its claws weren't very strong, so it probably didn't use them to defend itself or to fight unless it really had to. It was likely a peaceful creature, slowing gathering leaves to munch, and minding its own business.

Fast Facts

Name: *Therizinosaurus*
Size: 10 m (33 ft)
Lived: 70 million years ago, Mongolia
Ate: Plants

MONGOLIAN MONSTER

Therizinosaurus lived in Mongolia 75 million years ago and wasn't discovered until 1948. The dinosaur scientist who first found it thought the huge claws belonged to a giant turtle, 4.5 m (15 ft) long. It was another 25 years before it was recognized as a dinosaur.

I think you're clawsome.

THE SMALLEST DINOSAURS WERE THE SIZE OF SPARROWS

We tend to think of dinosaurs as huge, lumbering animals, and many of them *were* vast. But some were small, and even tiny. One that lived in Korea 110 million years ago was just the size of a sparrow, about 15 cm (6 in) long, and weighing only about 30 g (1 oz).

They said we should lighten up ...

FOLLOW THE FOOTPRINTS

The tiny dinosaur is known only from its footprints, just 1 cm (0.4 in) long. At first they look like bird footprints, but scientists can tell they are dinosaur footprints, as they show only two toes on the ground. The dinosaur was a little meat-eating raptor, and like other raptors, it held one toe off the ground, protecting its large "killing claw" (see page 38.)

Fast Facts

Name of tracks:
 Dromaeosauriformipes rarus
Size of footprint: 1 cm (0.4 in) long
Made: 110 million years ago, South Korea

MYSTERY TRACKS

Dinosaur footprints are common, but it can be hard to know what type of dinosaur made them. Sometimes there are bones nearby that give scientists a clue, but at other times there's no way of saying for certain. There's a lot we can tell from footprints, though, including the size and general type of the dinosaur—sauropod, theropod, or ornithischian. When scientists aren't sure, they name the footprint instead!

DINOS BIG AND SMALL

The largest dinosaurs were the titanosaurs—huge sauropods that could grow to 35 m (115 ft) long and weigh up to 80,000 kg (176,400 lb). That means the largest dinosaurs were about 230 times as long and weighed 2.7 million times as much as the tiny sparrow-dinosaurs.

DINOSAURS HAD HOLEY BONES

You'd think big, heavy dinosaurs needed big, heavy bones, solid and strong. But many dinosaurs had lots of holes in their bones. Being big and heavy takes a lot of energy. So although dinosaurs were big, their bodies tried not to be too heavy. Having holes in their bones helped to keep their weight down.

Not as heavy as I expected!

JUST BREATHE

When you breathe, muscles pull your diaphragm in your chest up and down, pulling air into your lungs and pushing your chest out. Birds—and probably dinosaurs—have different lungs that don't expand and contract like this. When a bird breathes in, air fills air sacs in its body, as well as its lungs, allowing oxygen to be absorbed into its bloodstream. When it breathes out, the air sacs and lungs both send out carbon dioxide.

HOLES HERE AND THERE

In a bird, there are large air sacs in the main part of its body. Dinosaurs had air sacs in their bones, often in the vertebrae (backbones), and extending into the neck and tail. We don't know much about how internal organs were arranged in dinosaurs, as they don't fossilize. But air sacs in the bones appear as holes.

LIGHT AND COOL

Having bones with lots of spaces in them kept large dinosaurs light. Birds need to be light to fly, but dinosaurs needed to be light so that their muscles didn't have to work too hard, which might have made them too hot. Sauropods were light for their size, compared with large mammals like elephants.

GIANT, FIERCE DINOSAURS WALKED ON TIPTOE

Huge dinosaurs like *Allosaurus* and *T. rex* tiptoed through the forests. Not because they were creeping up on animals they wanted to eat, but because that's just how their feet worked.

HEELS AND TOES

When you walk, you put your heel on the ground first, then your toes. Your footprints show that you place your whole foot on the ground. Theropod dinosaurs made footprints that show only toes and the front part of the foot. They didn't have heels, since the back part of their foot, near the ankle bones, couldn't touch the ground.

FURIOUS AND FAST

Theropods had to hunt their food, and their food would run away if it could. That meant they had to be fast to catch it. Animals that run on their toes generally go faster than those that put their whole foot on the ground. Fast mammals like cheetahs and wolves also run only on their toes (and fingers!).

FROM FOOT TO LEG

Theropods and other toe-walking animals do have all the normal bones of the leg and foot, but the foot is much longer in these animals than in similar animals that walk on their whole foot. The ankle and wrist are farther up the leg or arm.

Who has the best knees?

YOU COULDN'T HAVE LIVED IN THE DINOSAURS' WORLD ...

... and they couldn't have lived in yours. Forget *Jurassic Park*-style ideas of reviving dinosaurs. They would really struggle in the here and now. The world has changed, and now suits different types of animals, including us.

CHANGING CLIMATES

In the early days of the dinosaurs, the Triassic period, the world was warmer than it is now. The single supercontinent, Pangea, had very hot, dry summers and cold winters, with seasonal monsoons (heavy rains) near the coast. It was too warm for ice at the North and South Poles. That stayed much the same in the Jurassic and Cretaceous periods, with warm weather and high sea levels.

So, if we managed to recreate any dinosaurs today, they'd find it chilly.

CLIMATE PROBLEMS AREN'T NEW

We are rightly worried now about rising levels of carbon dioxide in the atmosphere causing global heating and climate change. But the dinosaurs saw even more catastrophic levels of change. Around 200 million years ago, volcanoes poured carbon dioxide into the air, causing climate change and mass extinction. There was about four times the level of carbon dioxide in the atmosphere than there is today. Until the early Jurassic period, there was also a lot less oxygen in the air than now. You wouldn't like it—it would be hard to breathe and far too hot.

T. REX ATE EACH OTHER ...

... Not as their preferred snack, but they would eat another *T. rex* if there was nothing better on the menu. Cannibalism—eating other animals of the same type—is quite uncommon. But *T. rex* was quite an unusual animal in many ways.

THE BITER BIT

Dinosaur scientists know that sometimes a *T. rex* would eat another *T. rex* because they have matched tooth marks on some *T. rex* bones to the type of marks made by *T. rex* teeth. Dinosaurs don't bite their own bones, so those marks mean another *T. rex* did the biting. *T. rex* was the only large predator living in the Hell Creek area where the bones were found, so it must have been the culprit.

Fast Facts

Name: *Tyrannosaurus rex*
Size: 12.5 m (41 ft)
Lived: 68–66 million years ago, northwest USA and western Canada
Ate: Other dinosaurs

TASTY BABIES

It's quite possible that a *T. rex* would have eaten another *T. rex*'s children, given the chance. There isn't room for a lot of large predators in any one area, so they would try to keep the competition down by eating rivals when they could—while they were still small. Although a big *T. rex* didn't have much to fear, growing up was a dangerous business. A *T. rex* would probably also eat a dead *T. rex* if it found one lying around. After all, why waste a good meal?

DYING DIDN'T DIE OUT

Although *T. rex* died out about 66 million years ago, some reptiles still alive today are cannibals. Komodo dragons and crocodiles both sometimes eat their own type. Again, babies are an easy option.

DINOSAURS THAT LIVED ON SMALL ISLANDS SHRANK

An island is a miniature habitat and some of the animals that live on islands become miniaturized.

FROM BIG TO SMALL

In North and South America, sauropods grew to be the largest animals that have ever walked on Earth. But on the islands that made up Europe at the time, there were mini-sauropods just the right size for a smaller habitat. *Magyarosaurus*, which lived in an area that's now part of Hungary, was only 6 m (20 ft) long and would come up to the shoulders of an adult human. But in South America, *Patagotitan* grew to 37 m (121 ft) long and weighed 57,000 kg (125,660 lb)!

ISLAND OF DWARF ANIMALS

Large animals becoming smaller over time if they are confined to an island is called "insular dwarfism." It's not only dinosaurs that got smaller. Later in time, European islands had dwarf hippos and elephants. In the last million years, a dwarf hippopotamus in Cyprus stood just 76 cm (2.5 ft) tall, which is half the size of a modern hippo.

> **Fast Facts**
>
> **Name**: Jurassic period
> **Date**: 200–145 million years ago
> **Temperature**: 5–10°C hotter than today
> **Atmosphere**: Carbon dioxide (0.16 per cent), oxygen (12 per cent)

ISLAND OF GIANTS

It can go the other way, too. Some small species can become much larger when they are stranded on islands. The island of Madagascar was home to the elephant bird, a giant flightless bird that grew to 3 m (10 ft) tall and died out a few hundred years ago. The Flores giant rat grows to more than 1 m (3.3 ft) including the tail—about twice as long as a brown rat.

DINOSAUR DROPPINGS WERE OFTEN HUGE

Not every fossil is a bit of a bone, tooth, or claw. There are also fossilized lumps of dino poop! Scientists can learn a lot from these about what the pooping dinosaur ate when it was alive.

They said I'd have to work my way up from the bottom ...

CUTTING UP COPROLITES

A lump of fossilized poop is called a coprolite. By looking inside it, scientists can work out what a dinosaur ate. The most famous coprolite is over 30 cm (12 in) long and weighs 7 kg (15.4 lb). It was made by a *T. rex* about 66 million years ago. It has chunks of bone in it, showing that *T. rex* ate quickly, crunching straight through bones and swallowing the pieces. That's an unusual way to eat. Most meat-eaters eat around the bones.

Fast Facts

Name: Coprolite
Found: 1824, by British fossil hunter Mary Anning
Made of: Whatever the animal ate

SAUROPOD SPLAT

Dinosaurs that ate plants would have made a different type of poop. Large sauropods that ate only leaves and twigs probably made large, soft heaps or pools of poop—perhaps like cow dung. They also made a lot of gas (see pages 92–93).

NEWER POOP

It's not just dino poop that scientists study. Looking at the fossilized poop of more recent animals, such as extinct giant sloths can help them to work out which types of plants they ate. This is done by finding chemicals in the coprolite that match those in plants living today. That's harder with dinosaur coprolites, since most of the plants and animals they ate no longer exist.

DINOSAURS ATE YOUR ANCESTORS

Not your great-granny, or even your great-great granny. Not any of your human ancestors. But if you go back far enough, your ancestors living at the time of the dinosaurs were tiny mammals—perfect for a dinosaur snack.

MAMMALS BIG AND SMALL

Humans are a type of mammal called a "primate." Other primates are monkeys and apes, including orangutans, gorillas, chimpanzees, spider monkeys, and lemurs. If you go back far enough in the human family tree, you get to the very first primates, about 60 million years ago. They looked a bit like a cross between a cat and a squirrel.

DINNER UNDERFOOT

Before the primates, and before the extinction of the dinosaurs, most mammals were small, and many were probably nocturnal (they came out at night). The first mammals evolved about 230 million years ago, alongside the dinosaurs. They would have been just the right size for a small dinosaur to eat, and for a big one to tread on. For these reasons, they probably lived in the trees, and maybe in burrows, keeping out of the way. It's likely that some were snaffled up for lunch by hungry dinosaurs, though.

Fast Facts

Name: *Psittacosaurus*
Size: 2 m (6.6 ft) long
Lived: 100 million years ago, China
Ate: Plants

Clear off! We dinos are meant to hunt mammals, not the other way around!

TURNING THE TABLES

Dinosaurs didn't always have it their own way. Fossils unearthed in China suggest that sometimes early mammals ate dinosaurs instead. One fossil preserved a mammal biting down on a small dinosaur, *Psittacosaurus*, and another fossilized mammal has bits of *Psittacosaurus* in its stomach.

SOME DINOS WERE GREAT PARENTS ...

... and some probably weren't! It seems that some dinosaur parents made a nice nest for their eggs, hung around until they hatched, and even brought food to their babies. Others did not.

LIFE IN THE NURSERY

A fossilized nest site in North America preserved *Maiasaura* nests, eggs, and babies from 77 million years ago. At least seven nests have been found in the same site, suggesting that *Maiasaura* nested in colonies. The fossils include hatchlings and older babies, so the babies stayed in the nest for a while. They couldn't walk immediately, so they needed their parents to bring them food and look after them.

Fast Facts

Name: *Maiasaura*
Size: 9 m (30 ft) long
Lived: 77 million years ago, northwestern USA and Canada
Ate: Plants

SITTING PRETTY

Maiasaura was a hadrosaur, a medium-sized, plant-eating dinosaur. *Maiasaura* didn't sit on their eggs. They were too heavy to do that. Instead, they probably covered their nests with leaves and other plant material. As this rotted, it produced heat, keeping the eggs warm. It also helped to keep them hidden.

Smaller, feathered dinosaurs did sit on their nests, spreading their feathers to keep the eggs warm and safe, as birds do now. A dinosaur from China, 70 million years ago, was found sitting over eggs containing embryos. So, feathers likely evolved to keep dinosaurs and their eggs warm, before they were used for flying.

THE ORIGINAL IGUANODON WASN'T AN IGUANODON

When someone discovers a new dinosaur, they get to choose a name for it. Usually, that name can't change, unless it's been accidentally given the same name as something else. But the dinosaur that scientist Gideon Mantell named *Iguanodon* in 1825 turned out not to be an *Iguanodon* at all. It's now named *Mantellisaurus*.

JUST TEETH

Gideon and Mary Mantell found fossilized teeth that looked quite like those of an iguana (a type of lizard) but much larger. Because of the resemblance, they assumed the teeth came from a giant lizard that looked like an iguana, and so named the animal *Iguanodon*. Nothing was known of the animal apart from the teeth, and it was another 20 years or so before the word "dinosaur" existed.

Let's call him Don, Iguano Don.

MINE, ALL MINE

In 1878, miners digging in Belgium found the remains of many *Iguanodon*, and for the first time, scientists put together a complete skeleton. These new fossils set the pattern for what *Iguanodon* looked like.

> **Fast Facts**
>
> **Name**: *Mantellisaurus*
> **Size**: 7 m (23 ft) long
> **Lived**: 125 million years ago, western Europe
> **Ate**: Plants

ISLAND FIND

In 1914, scientists found another dinosaur like *Iguanodon* on the Isle of Wight. It wasn't quite the same as the dinos from Belgium, so they named it *Mantellisaurus*. Then, it later became clear that the original *Iguanodon* teeth actually matched *Mantellisaurus* better than they matched those of the Belgian dinos. So, the first *Iguanodon* is a *Mantellisaurus* and a different dinosaur—the *Iguanodon* found in 1878—is now *Iguanodon*.

MICRORAPTOR HAD FOUR WINGS

Birds fly perfectly well with just two wings, so having four seems unnecessarily extravagant. Yet *Microraptor* had wings for arms, and also had wings on its legs.

You can never have too many wings!

ENJOYING THE HIGH LIFE

Microraptor was a small theropod dinosaur that lived in China 125 million years ago. Covered in feathers, and the size of a pigeon, it was one of the smallest dinosaurs, and—but for the extra wings—probably looked quite like a bird. Its four wings helped it glide between the trees where it lived, and possibly to fly properly by flapping.

Fast Facts

Name: *Microraptor*
Size: 90 cm (36 in) long
Lived: 120 million years ago, China
Ate: Small animals

AT HOME IN THE TREES

Microraptor's last meal was a kind of bird that lived in the trees. It had feet adapted to clutching branches, not to running on the ground or swimming. For *Microraptor* to have caught it, it must also have been good at moving through the trees, either by hopping or flying.

LESSONS FROM A LAST MEAL

Like other theropods, *Microraptor* was a meat-eater. One fossil revealed that *Microraptor* ate birds, since it had the bones of a bird in the area where its stomach would have been. From the bones, scientists could tell that the bird was head down, and not crunched up, so it had been swallowed whole. This means *Microraptor* probably caught it rather than finding it already dead, revealing that it was agile enough to catch birds.

A T. REX COULD HAVE WORN MITTENS

If you wear gloves, each finger and thumb has its own knitted sleeve. If you wear mittens, all your fingers are bunched together in one part and only your thumb is separate. A T. rex's gloves would always have been mittens because it had only two fingers!

Ready for a snowball fight?

COUNT THE FINGERS

Many animals have five fingers and toes, but dinosaurs often had fewer. All land-going animals with four limbs (quadrupeds) evolved from "fishapods"—fish-like animals that pulled themselves out of the sea around 380 million years ago. At least some fishapods had five bones in their front fins, which would evolve into fingers. It's easy for animals to lose fingers through evolution, but much harder for them to gain them, so most quadrupeds have five fingers or fewer.

Fast Facts

Name: *Tyrannosaurus rex*
Size: 12.5 m (41 ft)
Lived: 68–66 million years ago, northwest USA and western Canada
Ate: Other dinosaurs

GOING DOWN

You already know that many sauropods didn't have proper fingers (see pages 4–5), but they weren't the only dinos to cut down on them. As time went by, theropods gave up on fingers, too. Many, like *Allosaurus,* ended up with three "proper" fingers and toes, with the others shrinking or disappearing.

T. rex lost another finger, leaving only two. The small theropod *Mononykus* was extreme and kept just one finger on its hands. It possibly used its finger for poking termites out of holes so it could eat them.

MOST DINOSAURS WERE GROUNDED

The Jurassic skies were filled with flying reptiles, but they weren't dinosaurs. They were pterosaurs, which split from the dinosaur line before dinos became dinos. Dinosaurs ruled the ground, but pterosaurs ruled the sky. Some dinosaurs could at least glide, even if none (or few) could properly fly.

FALLING WITH STYLE

Animals that fly flap their wings, but animals that glide don't flap, and often don't even have wings. Instead, they might have flaps of skin stretched between their front and back limbs which work like a parachute. The flaps trap air beneath, holding the animal up as it moves between trees or rocks. To glide, an animal must be able to climb up somewhere to launch itself.

Fast Facts

Name: *Yi qi*
Size: About 80 cm (31 in) long
Lived: 159 million years ago, China
Ate: Possibly insects

FEATHERY BAT

The oddest gliding dinosaur was *Yi qi* (say "ee chee"). It probably looked like a cross between a bird and a bat, and was mostly covered with feathers. It was a theropod about the size of a large pigeon, with a long tail and a wing- or arm-span of about 60 cm (24 in).

MODERN GLIDERS

Animals that glide today include flying squirrels, sugar gliders, and flying foxes. They glide between trees, where they eat fruit. This saves a lot of climbing up and down! A particularly odd gliding animal is the flying frog. This has large webs between its fingers—enough to keep it in the air a bit longer than just jumping would.

TWO HUNDRED YEARS AGO, NO ONE KNEW DINOSAURS EVER EXISTED

The dinosaurs died out about 66 million years ago, and for most of the time since then they were forgotten, all traces of them gone. If you had been born a Roman or an ancient Egyptian, or even if you lived in the 1700s, you would never have known about dinosaurs.

BACK FROM THE GROUND

The first dinosaur bone recorded was found in Oxfordshire, England. Robert Plot described it in 1677, but decided it must be the end of the leg bone of a human giant. The first dinosaur fossils to be identified as relics of extinct animals were found in the 1820s. The word "dinosaur" was given to them by Richard Owen in 1842 when he realized they were the first examples of a group of huge, extinct reptiles.

This is no tall story.

DINOS AND DRAGONS

Dinosaur bones have been found for thousands of years, but people didn't know they were from extinct reptiles. In China, for instance, they were thought to be dragon bones that had magical properties. They were ground up and used in medicine. Drink your dinosaur juice—it's good for you!

Fast Facts

Name: *Megalosaurus* (first dinosaur found)
Size: 6–9 m (20–30 ft)
Lived: 166 million years ago, England
Ate: Other dinosaurs

This Jurassic juice is past its sell-by date.

THICK AND FAST

Soon after the first dinosaurs were recognized in England, others were found elsewhere in Europe, in North America, and in Africa. In the late 1800s and early 1900s, now-famous dinosaurs like *Diplodocus*, *Stegosaurus*, *T. rex*, and *Triceratops* were found. Imagine living without knowing about them!

PATAGOTITAN ATE 1,600 BOWLS OF SALAD A DAY

Probably not in bowls, though! The huge titanosaur had to eat 130 kg (290 lb) of leaves each day to stay alive. That's the equivalent of 1,600 bowls of salad.

EATING ALL DAY

Very large dinosaurs like *Patagotitan* had to spend most of the day eating. There wasn't much time for doing anything else—though we don't really know what else they would have liked to do with their time. The leaves and twigs they ate were low in nutrition, and didn't provide much energy, so they needed to eat a great deal to fuel their enormous bodies.

BIG BELLIES

Because plant matter is quite hard to digest, plant-eating dinosaurs had to have a very large belly to store all their food while it was slowly broken down. More food was always coming in, so there had to be a lot of space. If you look at sauropods—and also at chunky plant-eaters like *Ankylosaurus* and *Triceratops*—they have a large middle.

JURASSIC COWS

Today, plant-eating mammals like cows have special ways of digesting plants. They have more than one stomach, so one can start the process of breaking down the food and the next can take over for the following stage. Rabbits have another way—they partly digest their food, poop it out as blobs of soft jelly, and then eat it again. We don't know how dinosaur digestion worked, but their stomachs had a lot to cope with.

Fast Facts

Name: *Patagotitan*
Size: 31 m (102 ft) long
Lived: 102 million years ago, Argentina
Ate: Plants

"I don't know how your digestion works—but I'm not sticking around to find out!"

DEINOCHEIRUS WAS KNOWN ONLY FROM ITS ARMS FOR 50 YEARS

In the 1960s, fossil hunters found two huge arms, each ending in three fingers with enormous, fearsome claws. There was no clue as to the type of dinosaur they came from. Scientists just knew that its arms were 2.4 m (7.9 ft long), so it must have been big.

BITS OF DINOSAUR

It's quite rare to find all of a dinosaur in one go. Most skeletons are broken up before they fossilize, or only part is dug up or emerges from the rock. *Deinocheirus* remained a mystery for 50 years.

It might have looked like this!

Fast Facts

Name: *Deinocheirus*
Size: 11 m (36 ft) long
Lived: 70 million years ago, Mongolia
Ate: Water plants and soft land-growing plants

USE YOUR IMAGINATION

Dinosaur experts imagined different kinds of animal that the arms could have belonged to. The claws suggested it might be a fierce predator—perhaps like a *T. rex*, but one that used its arms to kill. Or maybe *Deinocheirus* used its claws like sloths do, to cling to trees. With no other parts of the body available, it was impossible to decide how the animal lived or where it fitted in with dinosaurs we know about.

PUTTING THE PIECES TOGETHER

Eventually, two more skeletons were uncovered, and *Deinocheirus* turned out to be really, really strange. It was huge—11 m (36 ft) long. Its back legs were stumpy and ended in hooves. It had a hump on its back and a duck-billed mouth. Fossils in its stomach suggested it ate plants and fish. It probably used its claws for digging up and pulling plants toward it from the boggy land where it lived.

ONE DINOSAUR LOVED SWIMMING

As far as we know, the only dinosaur that was enthusiastic about swimming was *Spinosaurus*, 95 million years ago. It had relatives that probably stood on river banks and snapped up fish, but *Spinosaurus* was dedicated to a watery way of life.

FISH SUPPER

Spinosaurus was fearsome and fierce. At 16 m (52.5 ft) long, it was bigger than a *T. rex*, with a huge sail on its back. But instead of stomping over the land hunting other dinosaurs, *Spinosaurus* spent its days swimming in rivers in North Africa and snatching fish—some as big as cars. Its crocodile-like mouth was filled with sharp, conical teeth, it had a long paddle-shaped tail perfectly adapted for swimming, and it might have had webbed toes.

PENGUIN BONES

Even the bones of *Spinosaurus* were adapted to life in the water. Like penguins and manatees, *Spinosaurus* had bones with thick walls. Their bones were dense, lacking the air holes that make the bones of most land creatures light. Heavy bones help animals stay underwater. Otherwise, they'd float up to the top and it would be hard to catch fish.

Fast Facts

Name: *Spinosaurus*
Size: 14 m (46 ft)
Lived: 95 million years ago, Egypt and Morocco
Ate: Fish and land animals or carrion

NO SOGGY SAUROPODS

People once thought that huge sauropods like *Diplodocus* spent most of their time in the water. The idea was that the water would help to hold up their heavy bodies and they could eat water plants. Now we know they lived on land, ate the leaves of trees and shrubs, and could hold their own bodies up perfectly well out of the water.

You need a swimming cap!

TITANOSAURS ABANDONED THEIR BABIES

Titanosaurs were truly huge dinosaurs, up to 37 m (121 ft) long, yet their babies were tiny. A fossil nest site in India shows that they made their nests in a colony and laid their eggs together, but the parents went off and left them all.

TOO BIG TO PARENT

A baby titanosaur was less than a thousandth the size of its mother and father. It had a lot of growing to do. But that size difference also meant that it would have been super easy for the parents to accidentally tread on their babies—or someone else's babies—especially in a crowded nest colony. The nests in the Indian colony are too close together for the titanosaur parents to have hung around. There just isn't room for them to have safely moved around between the nests.

YOU'RE ON YOUR OWN!

It seems parent sauropods scraped a hole in the soft mud of the nest site and laid their eggs in it, then left. They might have covered the eggs with foliage to keep them warm and hidden. The babies hatched with legs that were already strong enough to carry their weight, so they could probably run around immediately to look for food. Some sauropods might have made their nest sites at the edges of forests, too, so the babies could run into the trees to hide and feed in safety.

Fast Facts

Name: *Brontosaurus*
Size: 22 m (72 ft) long
Lived: 150 million years ago, North America
Ate: Low-growing plants

DINOSAUR FARTS WARMED THE WORLD

When gases such as carbon dioxide and methane collect in the air they make the world hotter. They are the cause of the climate crisis at the moment. But the Earth was hot in the time of the dinosaurs, too, and the methane in their farts and burps didn't help.

DINO GAS

Plant-eating animals like cows produce a lot of methane as they digest their food. They release this into the air from the top and the bottom. In today's world, humans keep so many cows that they are a major source of methane, adding to climate change. Giant sauropod dinosaurs would also have produced huge amounts of gas, including methane, as they digested leaves and other plant matter.

Fast Facts

Name: *Triceratops*
Size: 9 m (30 ft) long
Lived: 67 million years ago, western USA and Canada
Ate: Plants

INDUSTRIAL-SCALE GAS PRODUCTION

One scientific study suggested that sauropods would have added 472 million tonnes (520 million tons) of methane to the atmosphere every year. Today, human activities add about 360 million tonnes (397 million tons) of methane to the air each year, of which about a third comes from animal farming. That means dinosaurs could have been causing global heating at nearly four times the rate of modern cattle farming.

HERE AND GONE

Methane is 20 times more powerful as a greenhouse gas than carbon dioxide, so all those dinosaurs might have been giving themselves their own climate crisis. Methane doesn't stay in the atmosphere as long as carbon dioxide, though—it's gone in a few years. So when large plant-eating dinosaurs died out, the methane levels would have dropped quite quickly.

THE FIRST WHOLE DINOSAUR FOSSIL HID IN BOXES FOR 150 YEARS

Imagine lying buried underground for 193 million years, being dug up by an excited fossil hunter, but then just lying in a box for another 150 years. Sound depressing? It's exactly what happened to the first entire dinosaur fossil ever found.

TOP TO TAIL

Scelidosaurus was an early dinosaur covered in bony plates and spikes, rather like the later *Ankylosaurus* (but without the huge tail club). It lived in England in the early Jurassic period. The first bits of *Scelidosaurus* were found right at the beginning of the age of dinosaur hunting, in 1858. They were jumbled fragments from several animals.

Richard Owen (see page 82) persuaded the man who found the first bits to look for more. He soon uncovered a complete animal—the first time a complete dinosaur skeleton had ever been found. It was taken to London, where Owen looked at it quickly, wrote a few papers describing it, then ignored it. It stayed mostly ignored until the 2010s.

Fast Facts

Name: *Scelidosaurus*
Size: 4 m (13 ft) long
Lived: 190 million years ago, England
Ate: Plants

Last one to a T. rex pays for dinner!

BACK TO BOXES

Scelidosaurus wasn't unusual in being ignored for years and years. The American fossil hunters Othniel Charles Marsh and Edward Drinker Cope raced to dig up as many dinosaurs as possible in the 1870s and 1880s. But they dug them up—or had other people dig them up—faster than they could work on them. Crates and crates of fossils they recovered sat unopened for over a hundred years.

DINOSAURS WAGGED THEIR TAILS

Not like dogs, to show they wanted to be friends, but when they were running. It seems that some dinosaurs that ran on two legs, like the small but vicious *Coelophysis*, wagged their tails from side to side to help them go faster.

RUN ALONG, NOW

When you run, you pump your arms backward and forward as you move your legs. This is similar to the way some four-legged animals run, as they move both front and back legs to run. But your arms aren't on the ground, so why do you do it? If you pay attention next time you run, you'll notice that your left arm goes forward when your left leg goes backward, and vice versa. This helps you to balance. It also helps stop your torso (the central part of your body) moving too much and overbalancing you.

Fast Facts

Name: *Coelophysis*
Size: 3 m (10 ft) long
Lived: 215 million years ago, southwestern USA
Ate: Other dinosaurs

ARMS AND TAILS

Dinosaurs like *Coelophysis* had little arms. Moving them in time with its legs wouldn't have made much difference to the animal's balance. But it did have a hefty tail. Scientists recreating dinosaurs' running motion on computers found that swinging the tail from side to side could work in the same way as us moving our arms, helping the dinosaur to keep its balance as it ran.

AN ABELISAUR WAS VIRTUALLY ARMLESS

You might think *T. rex* had small arms, but that was nothing compared to the abelisaurs. Their arms were so short they were barely there. *Carnotaurus* was typical, with tiny arms that didn't bend.

Fast Facts
Name: *Carnotaurus*
Size: 8m (26 ft) long
Lived: 70 million years ago, Argentina
Ate: Other dinosaurs

ARMLESS, NOT HARMLESS

Carnotaurus was a meat-eater, 8 m (26 ft) long, with long back legs for chasing down its dinner—it's not clear what *Carnotaurus* ate. Some experts think it might have hunted down large sauropods, while others suspect it chased smaller animals. Whatever its preferred snack, it couldn't have used its arms to hold down its prey or bring any food to its mouth.

Just scratch under my chin.

I can't reach!

TINY ARMS THAT COULD NOT BEND

The front limbs didn't come close to matching the dinosaur's long, powerful back limbs. Its arms were "vestigial," which means that although the arms were still there, they didn't function like normal arms. The forearm (the bit between the elbow and the wrist) was only a quarter of the length of the upper arm. Your upper and lower arms are about the same length as each other.

USELESS FINGERS

The dinosaur had no wrist bones, so its finger bones attached straight to its lower arm. Its elbow didn't bend and it couldn't flex all its fingers. Only two of its four possible fingers had more than one bone, and the other two might just have been a bony growth. To make things even worse, the nerves going to the front limbs had largely wasted away over time.

ORYCTODROMEUS LIVED IN A BURROW

We think about dinosaurs stomping around over the landscape if they were large, or skittering through the undergrowth trying to escape being eaten if they were small. But fossils found in 2007 revealed that some dinosaurs made burrows and lived underground.

BURIED BABIES

A fossil of the small dinosaur *Oryctodromeus* was found in a burrow with two of its babies. The burrow was 30 cm (12 in) wide and 2 m (6.8 ft) long. An adult *Oryctodromeus* was also 2 m (6.8 ft) long, but it wouldn't have lain stretched out in its burrow. Its tail was probably quite flexible, so it could curl it around. Living in a burrow would have helped to protect it from predators and keep its babies safe. The temperature in a burrow is more stable than in the outside world, too, so the animals could have stayed comfortable in hot or cold seasons.

Fast Facts

Name: *Oryctodromeus*
Size: 2 m (6.8 ft) long
Lived: 95 million years ago, northwestern USA
Ate: Plants

DIG WITH YOUR NOSE!

Oryctodromeus had a body that was well adapted for burrowing. Its hips and shoulders were narrow, so it didn't need to make the burrow wide. That was useful, as a wide burrow could be big enough for predators to get into. The animal had a horny beak, which it might have used to dig through the soil. It doesn't sound very nice, digging with your nose and mouth, but it's worth it to avoid being eaten by a predator!

Can you dig an extra bedroom for the kids?

100 MILLION YEARS AGO, EVERY ANIMAL OVER 1 M (3.3 FT) LONG WAS A DINOSAUR

The dinosaurs started quite small, the underdogs in a world containing big, fierce crocodile-like creatures (see pages 52–53). But after a mass extinction 200 million years ago, they soon took over. Then, every big animal on land was a type of dinosaur.

ALL SHAPES AND SIZES

We think of dinosaurs as being big or even enormous. And this is true for some—the biggest sauropods could reach 35 m (115 ft). Nothing as large has ever lived on land before or since. But although all big animals were dinosaurs, not all dinosaurs were big. Some were the size of birds, some were giants, and plenty were somewhere in between.

Fast Facts

Name: *Liopleurodon*
Size: 6 m (20 ft) long
Lived: 160 million years ago, European seas
Ate: Animals

BIG IN THE SEA

Dinosaurs were the only large animals on land, but giant reptiles and huge fish swam in the sea. Plesiosaurs and pliosaurs could be 10 m (33 ft) long, and fish included a giant coelacanth that grew to 4.5 m (15 ft) long.

SMALL ON THE LAND

Animals on land that were less than 1 m (3.3 ft) long included lizards and other small reptiles, amphibians (like frogs and salamanders), birds, mammals, and different types of arthropod and insect. (Arthropods are animals with a hard outside and jointed bodies, such as scorpions.) Small and baby dinosaurs would have eaten all of these. But larger dinosaurs that wanted a bigger meal would have had to eat … other dinosaurs!

TRICERATOPS HAD 800 TEETH

You might expect meat-eating dinosaurs to have had more teeth than plant-eaters, but in fact plant-eaters had more. These grew in great tooth "batteries," with extra teeth stacked up ready for use.

READY AND WAITING

Tooth batteries are columns of piled-up teeth, closely packed in the dinosaur's jaw. Each column could have up to five teeth piled up. The teeth at the end of the column were used for chewing tough vegetation, which wore them away. When a tooth wore out, it dropped out and the next tooth in line took over. More teeth were always growing at the base of the column. In this way, *Triceratops* and other types of dinosaur were never short of teeth—they didn't have to live with a gap, like you do when you lose a baby tooth.

Fast Facts

Name: *Nigersaurus*
Size: 9 m (30 ft) long
Lived: 110 million years ago, Niger, Africa
Ate: Plants

BITING BEAK

If you came face to face with a *Triceratops*, you wouldn't see all those teeth in a toothy grin. The front of a *Triceratops*' mouth was a beak, made of horny material to snip through sticks and tough plants. Teeth farther back in the mouth chewed the plants. There were tooth batteries left and right, top and bottom, with each battery housing up to 200 teeth, about 40 of which would have been in use at any one time.

BATTERY POWERED

Lots of plant-eating dinosaurs had tooth batteries. Hadrosaurs, like *Parasaurolophus*, had them, and so did some sauropods, such as *Nigersaurus*. *Nigersaurus* worked like a lawn mower, with straight rows of narrow, stacked teeth that could shave off low-growing plants.

STEGOSAURUS DIDN'T HAVE ROOF TILES ON ITS BACK

But people once thought it did! The animal's discoverer, Othniel Marsh, at first thought *Stegosaurus* was a kind of water-going turtle with plates that overlapped like shingles on a roof.

Fast Facts

Name: *Stegosaurus*
Size: 7 m (23 ft) long
Lived: 150 million years ago, western USA and Portugal
Ate: Plants

PLATES AND SPIKES

Stegosaurs had arrangements of broad plates or narrow spikes along their backs. The spikes would probably have put other dinosaurs off attacking them. The spikes at the end of a *Stegosaurus* tail could give another dino a nasty whack. But the plates were too thin to provide much protection.

FANCY PLATES

In at least some species of *Stegosaurus*, the males had larger, rounder plates and the females had smaller, spikier plates. This suggests that the plates played a role in finding a mate. Males grew large, showy plates that would attract a female, while females grew spikier ones to scare away predators.

GETTING IT WRONG

Stegosaurus wasn't the only dinosaur people got wrong to start with. Early ideas of what dinosaurs looked like are often very different from modern ideas. *Iguanodon* had a thumb spike on its hands, but when this was first found, people assumed it went on the nose. And when reconstructing the marine reptile *Elasmosaurus*, its finder put the head on the wrong end!

It looks more like my Aunt Selma.

QIJIANGLONG WAS HALF NECK, HALF BODY

Can you imagine having a neck 9 m (30 ft) long? That's what the Chinese sauropod *Qijianglong* had. And its body was another 9 m (30 ft).

Qijianglong lived 150 million years ago in an area of forest with many rivers. Other dinosaurs of a similar type probably grew even larger, with necks maybe up to 16 m (52.5 ft) long. Strangely, these huge animals might have been able to rear up on their hind legs, and possibly even walk on them for a short distance.

Fast Facts

Name: *Qijianglong*
Size: 18 m (60 ft) long
Lived: 115 million years ago, China
Ate: Plants

TREETOP SNACKS

Dinosaurs with very long necks could reach high in the trees to eat leaves that weren't available to shorter animals. As dinosaurs needed a lot of food, they had to compete with each other to find enough. If a dinosaur had an advantage, it was more likely to survive. The only competition would have been with others of the same type.

Mind your toes!

CLOSE TO THE GROUND

In a forest, plants grow at all levels. While *Qijianglong* and other long-necked sauropods could feed on the highest leaves, different animals would have eaten plants growing close to the ground, such as ferns. In between, other dinosaurs could eat from low in the trees. And meat-eating dinosaurs of different sizes preyed on all these plant-eaters. A wide variety of animals made the most of the environment.

WE DON'T KNOW IF DINOSAURS WERE HOT OR COLD

Mammals (like you) are warm-blooded animals. They control their own body temperature no matter what the outside world is like. Reptiles like snakes and crocodiles are cold-blooded. They're slow and sluggish when the weather's cold and speed up when it's warm outside.

CROCODILES AND BIRDS

Crocodiles are reptiles, and are quite like dinosaurs. But birds are the only surviving dinosaurs, and they're warm-blooded. So were dinosaurs warm-blooded or cold-blooded? Scientists look at how dinosaurs lived and at their fossils to try to find out.

This weather is biting cold.

Fast Facts

Name: Warm-blooded mammals
Size: 2.5 cm (1 in) – 30 m (98 ft)
Eats: Animals and plants

HOT BLOOD

Warm-blooded animals use energy from food to keep their bodies at a steady temperature. They use more oxygen than cold-blooded animals, and their bodies work more quickly. They can be active in cold weather, or in the dark with no warming sunlight. A large body is easier to keep warm than a small body as it has less surface area compared to its overall size, and it's through their surface that animals lose heat.

Very large animals can overheat, and that's why animals like elephants aren't furry. But small animals can use fur or feathers to keep heat in. Many small dinosaurs had feathers or thin threads like hair which could keep them warm, but most big dinosaurs seem to have had scaly skin. Some, like *T. rex*, might have been fuzzy as babies and hairless later.

HOT OR COLD?

Warm- and cold-blooded animals typically have more or less oxygen in their bones. Dinosaurs fall into both groups. Those with less oxygen, typical of cold-blooded animals, include slow, plodding dinosaurs like *Triceratops* and *Stegosaurus*. Maybe some dinosaurs were cold-blooded and others were warm-blooded.

SPINOSAURUS WAS WIPED OUT IN WORLD WAR II

In fact, it was wiped out twice—once when it went extinct 93.5 million years ago, and again in 1944, when the only fossils of *Spinosaurus* were destroyed when the German museum where they were kept was destroyed in a British bombing raid.

LOST AND FOUND

Spinosaurus was not only lost twice, it was found twice. The first time was in 1912, when a few fossilized bones were found in Egypt by a German fossil hunter. Dinosaur scientist Ernst Stromer named and described the fossil in 1915 and made very careful drawings and notes. This is just as well, as his records became the only evidence that *Spinosaurus* had ever existed. Scientists worked from his notes and drawings for more than 60 years, until more bits of *Spinosaurus* were found in the 1990s.

PIECED TOGETHER

Dinosaurs are rarely found as whole individuals. More often, odd bones and teeth are found, separated from the rest of the body. There might be parts of several bodies mixed together, but still with lots of bits missing. Sometimes it's clear the bones are from more than one animal—if there are three arms or four skulls, for example.

But often it's not obvious until experts look more closely, comparing the size of different bones, for instance. What we know of *Spinosaurus* now has been pieced together from lots of finds. As more and more bits of *Spinosaurus* were found in the early 21st century, our ideas of what it looked like and how it lived changed, too.

IDEAS OF HOW SPINOSAURUS LOOKED

1910s–1960s

I might have had a small, bony crest.

1990s

2020s

Fast Facts

Name: *Spinosaurus*
Size: 14 m (46 ft)
Lived: 95 million years ago, Egypt and Morocco
Ate: Fish and land animals or carrion

KOSMOCERATOPS HAD THE FRILLIEST OF ALL FRILLS

Triceratops had a big neck frill, but some of its friends and relatives were even frillier. *Kosmoceratops* had the biggest frill and a total of 15 horns, far outdoing *Triceratops* with its three horns.

TOP TOPS ON CERATOPS

Ceratopsians are dinosaurs like *Triceratops* that had a head frill, horns, and a beaked mouth. They ate plants and were big, slow dinosaurs that walked on four legs. Some were quite dull, but others were truly spectacular. Top 'tops was probably *Kosmoceratops*, which had horns over each eye, one on its nose, a horn on each cheek, and ten horns that formed a fringe along the top of its frill. And it had little spikes along the edges of the frill. There's no beating that! Unless it's *Chasmosaurus*, or *Styracosaurus*, or *Achelousaurus* ...

THRILLING FRILLS

Lots of ceratopsian dinosaurs had spectacular arrangements of frills and horns, though a few were duller. Scientists aren't entirely sure what their ornaments were for. The horns and spikes probably helped to put off hungry attackers. A very large frill could perhaps have fooled predators into thinking the animal was even larger than it already was—something with a truly huge head.

The horn and frills might have attracted a mate, too. Today, animals such as stags have splendid, large antlers to attract female deer and also to tussle with rivals for those females. Perhaps female ceratopsians liked a particularly frilly male. A big frill might also have helped the animal to control its temperature, warming up in the sun and cooling down in the wind.

Fast Facts

Name: *Kosmoceratops*
Size: 4.5 m (15 ft) long
Lived: 76 million years ago, midwestern USA
Ate: Plants

ONLY 2.5 BILLION T. REX EVER LIVED

More or less! Scientists estimate there could have been as few as 140 million or as many as 42 billion, with 2.5 billion being the middle figure. They didn't all live at the same time, but were spread across 1.2–3.6 million years.

Fast Facts

Name: *Tyrannosaurus rex*
Size: 12.5 m (41 ft)
Lived: 68–66 million years ago, northwest USA and western Canada
Ate: Other dinosaurs

ONE IN A MILLION?

Very few of the animals that have ever lived have fossilized. There are only around one hundred *T. rex* fossils, and only 32 with enough bits remaining for us to know they are definitely from adults. That means even at the lowest estimate of 140 million *T. rex*, fewer than one in a million have survived (and been found) as a fossil.

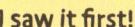

"I saw it first!"

FEW AND FAR BETWEEN

As huge predators that needed a lot of food, each *T. rex* required a large area to find its prey. *T. rex* only lived in the western part of North America, and there were probably around 20,000 alive at any one time. They were spread out. Somewhere the size of Washington, D.C. would have had only a couple of *T. rex*. Even somewhere as large as the whole of California could only support around 3,600 at a time. That gives each dinosaur 120 square km (46 square mi) to roam around.

ONE AT A TIME

T. rex was the apex (top) predator of its day. An environment can usually only support one or two types of top predator at a time. Similar dinosaurs, such as *Allosaurus* and *Carnotaurus*, lived at different times or in different places. *Allosaurus* lived in North America about 80 million years earlier than *T. rex*, and *Carnotaurus* lived at a similar time, but in Argentina.

STEGOSAURUS COULD BLUSH

Not on the cheeks, as you might blush, but on the plates along its back. But that doesn't mean *Stegosaurus* was easily embarrassed—we don't know about that. It had different reasons for making itself red.

BLOOD AND BONE

Stegosaurus's bony back plates, called "scutes," were covered with skin when the dinosaur was alive, and the skin held lots of blood vessels. By pumping more blood to its scutes, the dinosaur could make them flush red. There are two reasons it might have wanted to do this, and neither involves it being embarrassed.

Fast Facts

Name: *Stegosaurus*
Size: 7 m (23 ft) long
Lived: 150 million years ago, western USA and Portugal
Ate: Plants

CUTE SCUTES?

One reason is that the dinosaur used them in a mating display, trying to attract a partner. Perhaps red plates looked pretty to other *Stegosaurus*. In at least some species of *Stegosaurus*, the males had larger, rounder plates and the females had smaller, spikier plates. This sort of display is similar to the way birds show off their bright feathers, or do strange dances, to attract a mate. The shape and size of the scutes possibly also helped *Stegosaurus* to recognize each other as members of the same species, and so suitable as mates.

Stegosaurs know how to stay cool.

HOT AND COLD

Large scutes might also have helped *Stegosaurus* to control its temperature. By increasing blood flow to its scutes, the animal could have picked up heat from the environment or lost heat to it. It could stand with its scutes facing the sun to warm up, just like having solar panels. Or it could turn its scutes to the wind to cool down.

WE HAVEN'T FOUND ALL THE DINOSAURS YET

The first dinosaur fossils were found around 200 years ago, and from the middle of the 1800s more and more were uncovered. New dinosaurs are still discovered today, and more will be found in the future. Who knows what's waiting for us?

HERE, THERE, AND EVERYWHERE

The first dinosaurs were found in Europe, and soon after in North America. Since then, dinosaur fossils have been found in many parts of the world—Africa, South America, China, Mongolia, India, Australia, Greenland, and even Antarctica.

Many of the most famous dinosaurs have been found in the west of North America, but that doesn't mean most dinosaurs lived there. Fossils are preserved only if the conditions are right where an animal dies, and they're only uncovered if people look in the right places. Very many fossils have been found in China recently as fossil hunters have begun to look there.

Want to play with my dinosaur?

A DIFFERENT WORLD

The world of the dinosaurs wasn't like the world now. The climate was warmer and sea levels were higher. Some places that are now land were then under the ocean. If somewhere was under the ocean when the dinosaurs were alive, we won't find dinosaur fossils there. We don't look at land that is under forests and cities, so there could be fossils there, too.

FUTURE FINDS

So far, more than 1,000 types of dinosaur are known, and a new dinosaur is discovered roughly once a week. There must be a lot of surprises out there still!

PRONUNCIATION GUIDE

Abelisaur (ah-BEEL-ee-SORE
Achelosaurus (ah-KEL-oh-SORE-us)
Alamosaurus (ah-LAH-mow-SORE-us)
Allosaurus (AL-oh-SORE-us)
Ankylosaurus (an-KIH-loh-SORE-us)
Antarctopelta (ant-ARC-to-PELT-ah)
Archaeopteryx (ARK-ee-OPT-er-ix)

Borealopelta (bo-ree-AL-lo-pelta)
Brontosaurus (BRONT-uh-SORE-us)

Carnotaurus (KAR-no-TORE-us)
Cedarosaurus (SEE-dah-row-SORE-us)
Chasmosaurus (KAZ-moh-SORE-us)
Chixculub (CHICK-soo-loob)
Coelacanth (SEE-lah-can-th)
Coelophysis (SEE-loh-FISE-iss)
Cretaceous (cret-AY-shus)

Dakotaraptor (da-COAT-ah-rap-tuhr)
Diplodocus (dip-LOH-doh-kus)
Deinocheirus (DINE-oh-KIRE-us)

Elasmosaurus (el-LAZZ-moh-SORE-us)
Eoraptor (EE-oh-RAP-tuhr)

Hadrosaur (HAD-roh-SORE)
Herrerasaurus (HER-ray-rah-SORE-us)

Ichthyosaur (ICK-thee-oh-SORE)
Iguanodon (ig-WAH-noh-don)

Jurassic (joor-AH-sick)

Kosmoceratops (KOS-mo-SEH-ra-tops)

Liopleurodon (LY-oh-PLOO-ro-don)

Magyarosaurus (mag-YAR-oh-SORE-us)
Maiasaura (MY-ah-SORE-ah)
Mantellisaurus (man-TELL-ih-SORE-us)
Microraptor (MY-kroh-rap-tuhr)
Mononykus (mon-oh-NYE-kus)

Nigersaurus (NYE-jer-SORE-us)

Ornithischian (OR-ni-THISH-ee-an)
Oryctodromeus (oh-RICT-oh-DROME-ee-us)
Oviraptor (OH-vee-RAP-tuhr)

Pangea (pan-JEE-ah)
Parasaurolophus (PA-ra-sore-OL-off-us)
Patagotitan (pat-ah-GO-tie-tan)
Plesiosaur (PLEH-zee-oh-SORE)
Pliosaur (PLY-oh-SORE)
Prorotodactylus (pro-ROTE-oh-DACK-till-us)
pseudosuchian (SOOD-oh-SOOK-ee-an)
Psittacosaurus (SIT-ak-oh-SORE-us)
Pterosaur (TEH-roh-sore)

Qijianglong (CHEE-gee-ang-LONG)

Sauropod (SORE-oh-POD)
Scelidosaurus (SKEL-id-oh-SORE-us)
Sinosauropteryx (SIGH-no-sore-OP-tuh-rix)
Spinosaurus (SPINE-oh-SORE-us)
Stegosaurus (STEG-oh-SORE-us)
Styracosaurus (sty-RACK-oh-SORE-us)

Therizonosaurus (THEH-rih-ZIN-oh-sore-us)
Titanosaur (TY-tan-oh-SORE)
Triassic (try-AH-sick)
Tyrannosaurus rex (ty-RAN-oh-SORE-us REX)

Ubirajara (oob-ee-rah-ZHAR-ah)

Velociraptor (veh-LOSS-ee-rap-tuhr)
Yi qi (EE-CHEE)

Time periods:
Triassic: 252—201 million years ago
Jurassic: 201—145 million years ago
Cretaceous: 145—66 million years ago

123

GLOSSARY

ABELISAUR Large meat-eating dinosaur the same type of shape as T. rex but with even shorter arms.

ACID Corrosive substance that can dissolve hard materials.

AMPHIBIAN Animal that lays eggs in water but lives on land and breathes air as an adult.

ANKYLOSAUR Squat, plant-eating dinosaur covered with bony plates and spikes.

ANTARCTICA Continent over the South Pole.

ASTEROID Rock from space.

CELLULOSE Tough substance that forms the structure of plants.

COELACANTH Very old type of fish with fleshy lobs at the base of the fins.

CREST A fancy raised part on the head of an animal.

CRETACEOUS A period 145–66 million years ago.

CROCODILIAN An animal shaped like a crocodile.

CRUMHORN A J-shaped musical wind instrument.

CULPRIT Someone responsible for a crime or bad deed.

DEPOSITED Left somewhere

DESCENDANT Child, grandchild, or other person or animal in a direct line from an ancestor.

DIAPHRAGM A sheet of muscle that stretches across the inside of the chest, beneath the lungs.

EARTHQUAKE Shaking or lurching of the ground when large chunks of Earth's surface suddenly move.

EMBRYO The developing young of an animal, growing in an egg or the mother's body.

EVOLVED Developed over time to suit environmental conditions.

EXTINCTION EVENT An event that causes organisms to die out.

FOSSIL A relic of an organism preserved as stone. It can be part of the body or a trace such as a footprint or burrow.

FOSSILIZED Turned into a fossil by chemical changes.

HADROSAUR A medium-sized, plant-eating dinosaur.

HATCHLING A young animal that has just hatched from an egg.

HEMISPHERE Half of a sphere.

HIATUS A gap or pause.

INSULATION Material that stops the flow of heat, keeping something the same temperature.

JURASSIC A period 201–145 million years ago.

LANDLUBBER Someone who stays on land.

LANDSLIP A sudden fall of rocks or soil.

LARYNX An organ at the top of the throat used to produce sound.

MAMMAL An animal that has live young (rather than laying eggs), produces milk for its young and has at least some body hair or fur.

MASS EXTINCTION An event that wipes out a very large number of different kinds of organisms.

METHANE A gas made from hydrogen and carbon.

MICROORGANISM An organism too small to see without a microscope.

MINERAL A hard chemical substance with a crystal structure.

NOURISHMENT Wholesome food.

NUTRITION The process by which an organism uses food to support its life.

ORNITHISCHIAN A type of plant-eating dinosaur.

PANGEA A supercontinent that included all the dry land in a single block. It broke up around 250 million years ago.

PARASITE An organism that lives and feeds on another, such as a mosquito that sucks your blood.

PLESIOSAUR A prehistoric marine reptile with a long neck, small head, and four flippers.

PLIOSAUR A prehistoric marine reptile with a short neck, large head, and four flippers.

PREDATOR An animal that hunts and eats other animals.

PRESERVED Kept safely.

PREY An animal that is hunted and eaten by other animals.

PTEROSAUR A flying reptile.

QUILL A thin spine like the central shaft of a feather.

REPTILE A cold-blooded animal that hatches from an egg laid on land; breathes air and does not change form between hatching and adulthood.

REPTILIAN Like or related to reptiles.

SALAMANDER An amphibian that looks like a lizard.

SAUROPOD A plant-eating dinosaur with a large, heavy body, long neck and tail, and that walks on four legs. Diplodocus is a typical sauropod.

SEDIMENT Sand, mud, and small particles that collect in water.

SERRATED With a saw-like edge, divided into sharp teeth.

STREAMLINED With a smooth, tapered shape that moves easily through water or air.

SUPERCONTINENT A large landmass that includes most or all of the dry land on Earth.

TERMITE A small social insect that looks like an ant.

THEROPOD A fast, meat-eating dinosaur that walks upright on its large back legs. T. rex is a typical theropod.

TITANOSAUR A very large sauropod dinosaur.

TSUNAMI A large, flooding wave that washes over the coast causing destruction.

TUBA A brass musical instrument like a large trumpet.

VANE The central shaft of a feather to which the separate barbs are attached.

VEGETATION Growing plants.

VOLCANIC ERUPTION Scalding hot gases and molten rock pouring from a volcano.

VOLCANO A gap in the ground through which hot gases and rock from within Earth can pour out. Many volcanoes take the form of mountains, but they can also be a long crack in the ground.

INDEX

abelisaurs 98
Africa 83, 88, 120
air sacs 58–9
Alamosaurus 32–3
Allosaurus 60, 79, 117
ankylosaurs 51
Ankylosaurus 84, 94
Antarctica 20, 44–5, 120
Antarctopelta 44
Archaeopteryx 9
Argentina 51, 85, 98, 117
arms 86, 96–7, 98–9
vestigial 99
arthropods 103
asteroid strikes 32, 34–5
Atlantic Ocean 53

balance 96–7
beaks 37, 48, 51, 54, 101, 105, 114
Belgium 75
birds 8–9, 14–17, 19, 22, 28–9, 34, 37, 43, 47–8, 58–9, 67, 73, 76–7, 103, 110, 119
body shape 51
body temperature 110–11, 115, 119
bones
holey 58–9, 89
see also wishbone (furcula)

Borealopelta 49
Brazil 42–3
breathing 58–9
Brontosaurus 32, 91
burrows 27, 71, 100–1

California 30, 117
Canada 11, 17, 25, 49, 64, 72, 78, 92, 105, 116
cannibalism 64–5
carbon dioxide 45, 50, 58, 63, 67, 92–3
Carnotaurus 98–9, 117
Cedarosaurus 23
ceratopsians 114–15
Chicxulub asteroid 35
China 7–9, 71, 73, 76–7, 80, 83, 108, 120
claws 5, 9, 38, 54–5, 86–7, 99
killing 38, 56
climate change 35, 45, 50, 62–3, 92–3, 121
coelacanths 103
Coelophysis 96–7
cold-blooded animals 110–11
competition 52, 65, 109
Cope, Edward Drinker 95
coprolites 68–9

Cretaceous period 32, 45, 62
crocodiles 7, 16, 53, 110
crocodilians 52–3, 102

Dakotaraptor 28–9
Deinocheirus 86–7
digestion 84–5
dinosauromorphs 50–1
Diplodocus 4–5, 27, 32, 41, 83, 89
dwarfism, insular 67

eggs 5, 14, 46–7, 72–3, 90–1
Egypt 89, 112–13
Elasmosaurus 107
elephant bird 47, 67
England 82–3, 94–5
Europe 32, 33, 66, 75, 120
European seas 102
extinction 8, 34–5, 63, 71, 82–3, 112
extinction events 32–5, 50, 102

farts 92–3
feathers 6–9, 14, 27, 36–7, 40, 43, 73, 76, 111, 119
feet 4–5, 60–1
fingers 78–9, 99
fish 31, 34, 87, 88, 88–9, 103, 113

fishapods 78
fleas 36–7
flight 6, 9, 28–9, 59, 73, 76, 80
fluorescence 48–9
footprints, fossilized 56–7, 60
fossils 12–13, 26–7, 30–1, 33–4, 36, 44, 71, 74–5, 77, 82, 90, 110, 112, 116
first complete 94–5
fluorescent 48–9
footprints 56–7, 60
fossilization process 12–13, 26–7
future finds 120–1
incomplete 40–1, 86–7, 113
poo (coprolites) 68–9
shapes 27
trace fossils 27
tracks 5
furcula (wishbone) 28–9

gaseous exchange 58
gastroliths 22–3
Germany 9
gigantism 67
gliding animals 80–1
global heating 63, 92–3
Great Dying 50, 52
greenhouse gases 93

126

see also carbon dioxide; methane
growth 24–5

hadrosaurs 51, 73, 105
hands 78, 107
hearing 17
Hell Creek 64
Herrerasaurus 51
hippopotamus, dwarf 67
horns 114–15
human ancestors 70–1
Hungary 66

Iguanodon 74–5, 107
India 90, 120
infrasound 17
insects 34, 36, 80, 103
island life 66–7
Isle of Wight 75

Jurassic period 32–3, 50, 62–3, 67, 80, 85, 94
Jurassic pseudo-flea 37

keratin 42, 49
Komodo dragon 65
Korea 56–7
Kosmoceratops 114–15

legs 61, 96–9
"leks" 14
Liopleurodon 102

Madagascar 47, 67
Magyarosaurus 66
Maiasaura 72–3
mammals 19, 39, 59, 61, 70–1, 85, 103, 110
Mantellisaurus 74–5
Marsh, Othniel Charles 95, 106
mass extinction 50, 63, 102
mating 14–15, 43, 48, 107, 115, 119
meat-eaters 7–9, 17, 19, 21, 25, 29, 35, 39, 43, 51, 64–5, 68, 77–8, 83, 89, 97–8, 102–4, 109, 113, 116–17
Megalosaurus 83
methane 92–3
Mexico 34
Microraptor 76–7
miniaturization 66–7
Mongolia 21, 29, 39, 41, 47, 55, 86, 120
Mononykus 79
Morocco 89, 113

nails 38, 42
naming dinosaurs 74
neck frills 114–15
necks 108–9
nests 5, 6, 14, 46, 72–3, 90–1
Nigersaurus 105
nocturnal animals 39, 71

North America 10, 18, 20, 30–1, 32–3, 66, 83, 91, 95, 117, 120
North Pole 45, 62

omnivores 47, 110
ornithischians 51, 57
Oryctodromeus 100–1
Oviraptor 47
Owen, Richard 82, 95
oxygen 58, 67, 111

Pangea 21, 31, 53, 62
Parasaurolophus 10–11, 105
parasites 36–7
parenting 72–3, 90–1
Patagotitan 66, 84–5
plant-eaters 5, 11, 19, 22–3, 33, 35, 41, 44, 51, 54–5, 69, 71–3, 75, 86, 91–3, 95, 101, 104–6, 108–9, 114–15, 118
digestion 84–5
teeth 104–5
plate tectonics 44

plates ("scutes") 106–7, 118–19
plesiosaurs 31, 103
pliosaurs 103
poo, fossilized 68–9
Portugal 19, 106, 118
predators 49, 64–5, 87, 100–1, 107, 115
apex 117
primates 70–1
Prorotodactylus 50
pseudosuchians 52–3
Psittacosaurus 71
pterosaurs 7, 80

Qijianglong 108–9
quadrupeds 78
quills 6–7

raptors 8, 15, 20–1, 28–9, 38–9, 47, 56, 76–7
rat, Flores giant 67
reptiles 6–7, 9, 19, 22, 45, 65, 80, 82–3, 103, 107, 110
"ribbons" 42–3
roaring 16–17

"sauropod hiatus" 32–3

127

sauropods 4–5, 32–3, 37, 46, 51, 57, 59, 66, 69, 79, 84, 89, 91, 92, 98, 102, 108–9
Scelidosaurus 94–5
"scutes" (plates) 106–7, 118–19
sea levels 45, 62, 121
Second World War 112
sharks 31
Sinosauropteryx 7
size of dinosaurs 50–1, 52, 56–7, 66–7, 102–3, 108–9
soundscape 10–11, 16–17
South America 20, 33, 51, 66, 120
South Pole 45, 62
speed 61
spikes 106–7, 114–15, 119
Spinosaurus 88–9, 112–13
Stegosaurus 18–19, 83, 106–7, 111, 118–19
stone-eating 22–3
Stromer, Ernst 112
swimming dinosaurs 88–9

tails 106
wagging 96–7

teeth 9, 22, 26, 37, 40, 52, 64, 74–5, 88, 104–5, 113
Therizinosaurus 41, 54–5
theropods 8–9, 14, 29, 38–9, 51, 57, 60–1, 76–7, 79, 81
ticks 36–7
titanosaurs 57, 90
toes 60–1, 78, 79, 88
tooth batteries 104–5
trace fossils 27
tracks, fossilized 5
Triassic 62
Triceratops 27, 83, 84, 92, 104–5, 111, 114
tsunami 35
Tyrannosaurus rex 8, 15–18, 20–1, 24–5, 27, 60, 64–5, 68, 78–9, 83, 87–8, 98, 111, 116–17

Ubirajara 42–3
ultraviolet (UV) 48–9
United States 5, 11, 17, 19, 23, 25, 33, 39, 64, 72, 78, 92, 97, 101, 105–6, 115–16, 118
Velociraptor 8, 15, 20–1, 28–9, 38–9
volcanic activity 50, 53, 63

warm-blooded animals 110–11
Western Interior Seaway 30–1, 33
wings 8–9, 28, 76–7, 80
wishbone (furcula) 28–9

Yi qi 80–1